T0212289

Understanding Atrial Fibrillation:
The Signal Processing Contribution
Part I

SYNTHESIS LECTURES ON BIOMEDICAL ENGINEERING

Editor
John D. Enderle, *University of Connecticut*

Understanding Atrial Fibrillation: The Signal Processing Contribution, Part I
Luca Mainardi, Leif Sörnmo, and Sergio Cerutti
2008

Lung Sounds: An Advanced Signal Processing Perspective
Leontios J. Hadjileontiadis
2008

An Outline of Information Genetics
Gérard Battail
2008

Neural Interfacing: Forging the Human-Machine Connection
Thomas D. Coates, Jr.
2008

Quantitative Neurophysiology
Joseph V. Tranquillo
2008

Tremor: From Pathogenesis to Treatment
Giuliana Grimaldi and Mario Manto
2008

Introduction to Continuum Biomechanics
Kyriacos A. Athanasiou and Roman M. Natoli
2008

The Effects of Hypergravity and Microgravity on Biomedical Experiments
Thais Russomano, Gustavo Dalmarco, and Felipe Prehn Falcão
2008

Understanding Atrial Fibrillation: The Signal Processing Contribution, Part I

Luca Mainardi, Leif Sörnmo, and Sergio Cerutti

ISBN: 978-3-031-00505-3   paperback
ISBN: 978-3-031-01633-2   ebook

DOI 10.1007/978-3-031-01633-2

A Publication in the Springer series
SYNTHESIS LECTURES ON BIOMEDICAL ENGINEERING

Lectures #24
Series Editor: John D. Enderle, *University of Connecticut*

Series ISSN
Synthesis Lectures on Biomedical Engineering
Print 1930-0328   Electronic 1930-0336

Understanding Atrial Fibrillation:
The Signal Processing
Contribution
Part I

Luca Mainardi
Department of Bioengineering
Politecnico di Milano, Italy

Leif Sörnmo
Department of Electrical Engineering
Lund University, Sweden

Sergio Cerutti
Department of Bioengineering
Politecnico di Milano, Italy

SYNTHESIS LECTURES ON BIOMEDICAL ENGINEERING #24

ABSTRACT

The book presents recent advances in signal processing techniques for modeling, analysis, and understanding of the heart's electrical activity during atrial fibrillation. This arrhythmia is the most commonly encountered in clinical practice and its complex and metamorphic nature represents a challenging problem for clinicians, engineers, and scientists. Research on atrial fibrillation has stimulated the development of a wide range of signal processing tools to better understand the mechanisms ruling its initiation, maintenance, and termination. This book provides undergraduate and graduate students, as well as researchers and practicing engineers, with an overview of techniques, including time domain techniques for atrial wave extraction, time–frequency analysis for exploring wave dynamics, and nonlinear techniques to characterize the ventricular response and the organization of atrial activity. The book includes an introductory chapter about atrial fibrillation and its mechanisms, treatment, and management. The successive chapters are dedicated to the analysis of atrial signals recorded on the body surface and to the quantification of ventricular response. The rest of the book explores techniques to characterize endo- and epicardial recordings and to model atrial conduction. Under the appearance of being a monothematic book on atrial fibrillation, the reader will not only recognize common problems of biomedical signal processing but also discover that analysis of atrial fibrillation is a unique challenge for developing and testing novel signal processing tools.

KEYWORDS

Atrial fibrillation, signal processing, signal modeling, atrial activity extraction, spectral analysis, time–frequency analysis, ventricular response characterization, heart rate variability, atrial organization indices, source modeling, volume conductor modeling, intracardiac AF detection, AF long-term monitoring, implantable cardiac devices pacemakers.

AUTHORS

Andreas Bollmann
Department of Electrophysiology, Leipzig University Heart Center, Leipzig, Germany

Andreu Climent
Electronic Engineering Department, Valencia University of Technology, Valencia, Spain

Valentina Corino
Department of Bioengineering, Politecnico di Milano, Milano, Italy

Luca Faes
Department of Physics, University of Trento, Trento, Italy

Jeff Gillberg
Medtronic Inc., Minneapolis, Minnesota, USA

Vincent Jacquemet
Department of Cardiology, University of Lausanne, Lausanne, Switzerland

Federico Lombardi
Division of Cardiology, San Paolo Hospital, University of Milan, Milan, Italy

Luca Mainardi
Department of Bioengineering, Politecnico di Milano, Milano, Italy

Rahul Mehra
Medtronic Inc., Minneapolis, Minnesota, USA

Adriaan van Oosterom
Department of Cardiology, University of Lausanne, Lausanne, Switzerland

Simona Petrutiu
Department of Electrical Engineering, Northwestern University, Chicago, USA

Flavia Ravelli
Department of Physics, University of Trento, Trento, Italy

José Joaquín Rieta
Biomedical Synergy, Valencia University of Technology, Gandia, Spain

Alan Sahakian
Department of Electrical Engineering, Northwestern University, Chicago, USA

Frida Sandberg
Department of Electrical Engineering, Lund University, Lund, Sweden

Shantanu Sarkar
Medtronic Inc., Minneapolis, Minnesota, USA

Leif Sörnmo
Department of Electrical Engineering, Lund University, Lund, Sweden

Martin Stridh
Department of Electrical Engineering, Lund University, Lund, Sweden

Steven Swiryn
Division of Cardiology, Northwestern University, Chicago, USA

Paul Ziegler
Medtronic Inc., Minneapolis, Minnesota, USA

Contents

Preface

Atrial fibrillation (AF) is a widely diffused arrhythmia which affects 2 million people in Europe and roughly 2.2 million in the U.S., reducing quality of life and increasing risk for stroke and death. Atrial fibrillation is the most common cause of hospitalization and its treatment is one of the most cogent issues in clinical arrhythmology. Considerable research effort is currently directed to this arrhythmia because the mechanisms causing its initiation, maintenance, and termination are not sufficiently well understood. In addition, as AF prevalence and incidence doubles with each decade beyond 50 years of age, the impact of this arrhythmia will be progressively larger in the near future due to the aging population.

The complex, heterogeneous, and metamorphic nature of this arrhythmia represent a challenging problem for clinicians, engineers, and scientists, stimulating the development of quantitative methods for the analysis of the electrical signals recorded during AF. These are actually considerably larger than those employed to analyze any other cardiac arrhythmias.

Nevertheless, ECG-based analysis of AF has, for a long time, been confined to a qualitative characterization of the fibrillatory waves. Only in the last decade has more detailed information contained in the fibrillatory waves emerged, leading to more detailed characterization of this arrhythmia. This progress is largely due to the advent of reliable signal processing procedures which are able to separate atrial and ventricular activities.

This book presents a comprehensive overview of signal processing methods employed for modeling, analysis, and management of the heart's electrical activity during AF. While some signal processing challenges are common to those encountered in the analysis of other bioelectrical signals, other challenges are unique to the analysis of AF. The topics are related to spectral analysis (traditional and time-varying), noise-reduction, separation of signal sources, and assessment of non-linear dynamics. We are therefore confident that this book will be of interest to researchers which are familiar with biomedical signal processing and to students who wants to learn about the new applications and challenges of biomedical signal processing. In addition, our hope is that this book will serve as an inspiration to further research on AF—a research area which is still in need of much exploration.

CONTENTS OVERVIEW

Chapter 1 presents current clinical knowledge about human AF. Starting from electrophysiological mechanisms and determinants of AF, the chapter considers AF incidence and prevalence and AF classification schemes used in clinical practice. The clinical consequences of AF as well as various therapeutic options are also described.

Chapter 2 describes time domain methods designed to characterize the atrial fibrillatory waveforms in ECG. Time domain parameters such as amplitude and rate are considered as well as their reproducibility over time. The value of vectorcardiographic analysis of f-waves is discussed at length as is the choice of lead configurations which are particularly suitable for analysis of AF.

Chapter 3 addresses with the fundamental problem of how to extract the atrial fibrillatory activity from the ECG. Different approaches for atrial extraction are reviewed, ranging from basic average beat subtraction to principal and independent component analysis. The problem of how to evaluate and compare the performance of different algorithms is considered.

Chapter 4 deals with the estimation of AF frequency, i.e., the repetition rate of the f-waves. This parameter has been found valuable in various clinical applications. Different techniques for time–frequency analysis of the atrial signal are presented, designed to unveil temporal variations in AF frequency which are either spontaneous in nature or due to some kind of intervention.

Chapter 5 summarizes methods for analysis of the ventricular response during AF and description of AV conduction characteristics. Such analysis is based on the description of RR interval variability using various signal processing tools of increasing complexity: from simple statistical indices to more sophisticated metrics derived from nonlinear dynamics theory.

In Chapter 6, the focus is shifted from the body surface to endo- and epicardial recordings. The problem of assessing "atrial organization" is introduced and as this term assumes different meanings, particular attention is paid to the peculiar characteristics of organization captured by each method. These characteristics include repeatability/regularity of the atrial activations, correlation/synchronicity among electrograms recorded in different locations, and similarity of the atrial wave morphology.

Chapter 7 introduces a computer model suitable for the interpretation and analysis of fibrillatory ECGs. Starting from a biophysical model of atrial cell membrane during AF, the basics of the forward problem are considered and used to generate simulated ECG signals which mimic the ones observed on the thorax. The model based approach provides a link between body surface recordings and atrial activity and puts into evidence the information content of these signals.

Chapter 8 deals with the problem of detecting atrial tachycardia/atrial fibrillation in implantable device, being crucial for long-term monitoring in which high specificity is usually required to avoid over-treatment of nonsustained atrial arrhythmias. After having introduced the clinical requirements of such a monitoring device, the chapter presents the details of a detector for atrial tachyarrhythmias which is based on the analysis of RR interval irregularity.

ACKNOWLEDGEMENTS

This book was made possible because of the collaboration between scientists who have made contributions to atrial fibrillation research. We wish to express our sincere gratitude to all the contributors for generously sharing their expertise. Without their great enthusiasm and willingness to spend precious time on writing, this book would never have happened.

We are also grateful to Dr. Sih who provided material for replicating Fig. 6.1 in Chapter 6.

Luca Mainardi
Leif Sörnmo
Sergio Cerutti
December 2008

Part I

CHAPTER 1

Introduction to Atrial Fibrillation: From Mechanisms to Treatment

Andreas Bollmann and Federico Lombardi

1.1 BACKGROUND

Atrial fibrillation (AF) is a supraventricular arrhythmia characterized by chaotic and uncoordinated atrial activation and contraction. It represents a major clinical, social, and economical burden, and its importance is expected to increase even more in the near future. The progressive aging of the general population and more precise diagnostic capabilities are associated with an inevitable rising incidence and prevalence of this rhythm disorder, which limits functional capability, favors occurrence of cerebrovascular events, and increases people's request for emergency room visits and hospital recovery. Moreover, several new therapeutical targets for AF are progressively affirming. This notion is supported by the fact that papers concerning AF, and indexed in www.pubmed.com during the last 15 years, have progressively increased by about 7 times, i.e., 4 times more than any other clinical argument on the web in the same period of time [1].

Atrial fibrillation has several heterogeneous clinical presentations, starting from paroxysmal forms in which the only medical trouble is its sudden onset to other types in which AF is merely a clinical manifestation of underlying cardiac or noncardiac pathologies (e.g., pulmonary, nephrologic, or endocrine diseases).

The wide heterogeneity of its underlying pathophysiology, its different symptomatic impact on each patient, and the possibility of different therapeutical options represent nowadays a great challenge for clinicians and also for several national health systems.

In this chapter, electrophysiological mechanisms, AF classification schemes, its incidence and prevalence, clinical consequences, as well as therapeutic options will be described.

1.2   ATRIAL FIBRILLATION MECHANISMS

Atrial fibrillation has a complex and not completely understood pathophysiology. Genetic predisposition, structural changes and fibrosis, progression of heart disease, inflammation, autonomic dysfunction coupled with electrophysiological abnormalities of the atria, and pulmonary vein (PV) sleeves, may all act, to various degrees, as contributors to initiation and maintenance of the fibrillatory process.

Early experimental studies and computer models suggested that AF is due to multiple reentrant wavelets. With evolving mapping technologies applied during open heart surgery or electrophysiologic studies, stable or unstable reentrant circuits with short cycle length in or near the PVs have been identified as AF drivers, while the other atrial parts may follow passively with fibrillatory conduction.

As electrophysiological mechanisms of human AF are the focus of this chapter, findings restricted to AF activation mapping and AF frequency mapping will be reviewed. Before that, the wavelength concept is introduced which seems crucial for the understanding of atrial activation in general and reentry mechanisms in particular.

1.2.1   MAIN DETERMINANTS OF ATRIAL ACTIVATION – THE WAVELENGTH CONCEPT

The average size of reentry pathways during AF is dependent on atrial wavelength, defined as the product of conduction velocity (CV) and refractory period, i.e., the time that atrial tissue is not excitable [2]. Long wavelengths are associated with larger and fewer wavefronts, while short wavelengths result in a larger number of smaller circuits which may have implications for arrhythmia susceptibility and stability as well as for therapeutic effects of antiarrhythmic drugs, cardioversion, or ablation.

Several groups have developed animal models of AF in order to assess structural, functional, and electrophysiological requirements for AF induction and sustenance. One common finding was progressive shortening of atrial refractoriness in models involving chronic rapid pacing [3] or electrically maintained AF [4, 5]. Previous experimental studies have also shown that long-term rapid atrial pacing induces atrial conduction slowing [3, 6, 7].

The findings in human studies are in close agreement with those obtained in experimental AF models. Daoud et al. investigated the effective refractory period (ERP) changes in pacing-induced AF [8, 9]. They found a significant ERP shortening (up to 20%) after a few minutes. Similar observations have been reported by Yu et al. [10]. The authors also showed that ERP shortening is a rate-dependent process with shorter cycle length, leading to a greater decrease in ERP. That ERP shortening is an essential feature in AF is further supported by several studies [11, 12, 13, 14, 15, 16, 17], measuring these parameters in AF patients after cardioversion and comparing them with normal controls. Interestingly, ERP changes have been shown to be reversible in nature [14, 17, 18], and are the result of the high and irregular rate. Normalization of ERP shortening was observed as early as 12–24 h following cardioversion [17, 18].

While the time course of CV changes associated with persistent AF in humans is not accessible, the recovery from these changes was studied in several investigations [14, 17, 19]. In fact, they mirrored the time course of the electrophysiological changes (fast ERP shortening, slow CV depression) found in experimental AF models. While ERP returned to normal values within days following cardioversion of persistent AF [14, 17], P wave duration reached control levels within one month when obtained from the surface ECG [17], but began to recover only after 6 months when assessed by the P wave signal averaged ECG. The delayed recovery may indicate a severely disturbed electrical atrial function and seems to be responsible for the clustering of AF recurrences early after cardioversion [20].

1.2.2 AF ACTIVATION MAPPING

One widely accepted theory of AF mechanisms was proposed by Moe as early as in 1962 [21]. It postulated that AF perpetuation is based on the continuous propagation of multiple wavelets wandering throughout the atria. In 1985, mapping of experimentally induced AF in canine hearts provided the first evidence supporting Moe's multiple wavelet hypothesis [22]. Mapping of the atrial electrical activation pattern during human AF has further improved our understanding of the fibrillatory process supporting in part, but also challenging, the multiple wavelet hypothesis.

Konings et al. found different activation patterns in induced AF in patients with the Wolff–Parkinson–White syndrome undergoing cardiac surgery, using epicardial high density mapping of the right atrial free wall [23]. They categorized the arrhythmia according to the number and complexity of activation wavefronts detectable in small atrial areas (3.6 cm diameter). Type I AF was characterized by a single uniform wavefront, type II AF showed one nonuniform or two simultaneous wavefronts, and in the more complex type III AF multiple wavefronts were present simultaneously (Figure 1.1). Similarly, Holm et al. were also able to identify different atrial activation patterns by applying high-resolution epicardial mapping of the right atrial free wall to patients with persistent AF [24]. They classified activation as:

1. unorganized activation with several simultaneously present activation waves (inconsistent pattern, type III);

2. predominantly organized activation with either frequent episodes of uniform activation or frequent episodes of activation with focal spread (consistent pattern, types I and II); and

3. focal preferable activation.

While the aforementioned mapping studies were restricted to the right atrium, other studies of permanent AF in patients undergoing open heart surgery, using multielectrode strips placed on small areas of both atria, recorded localized areas with rhythms of short and either irregular or regular cycle lengths [25, 26, 27]. This finding is consistent with the concept that AF, in certain cases, is maintained by a driver and passive fibrillatory conduction to other atrial parts; this finding is also supported by AF frequency mapping.

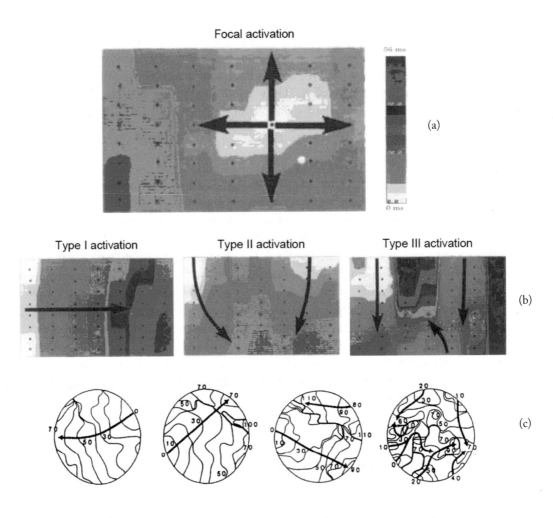

Figure 1.1: Activation maps with isochrones of the right atrial wall reveal different AF activation types. (a) Focal activation and (b) type I, II, and III activation with increasing number of simultaneous waves and, consequently, increasing complexity of atrial activation (modified from [24]). (c) In type I, just one wavefront with homogeneous wavefront propagation is present. Type II refers to either one wavefront with inhomogeneous wavefront propagation (left) or two simultaneous wavefronts (right). In type III with the most complex activation, multiple (at least more than two) wavefronts are present at the same time (modified from [23]). Arrows indicate propagation directions of individual wavelets.

1.2.3   AF FREQUENCY MAPPING

Time domain analysis of atrial activation is difficult to interpret, whereas frequency mapping using spectral analysis techniques allows simpler analysis of AF activity and enables localization of sites of rapid and periodic activity that likely are closely linked to sources maintaining AF.

Recent atrial mapping studies suggest that—depending on the patient population investigated—various frequency gradients within the atria/PV can be found [28, 29]. Lazar et al. noted frequency gradients in patients with paroxysmal AF with the highest fibrillatory frequencies in the left atrium/PV, intermediate in the coronary sinus (CS), and lowest in the right atrium (RA). However, fibrillatory frequencies were similar among different sites in patients with persistent AF [29]. Moreover, within a different patient group with paroxysmal AF, Lin et al. found that the location of the highest fibrillatory frequency was dependent on the arrhythmogenic PV or superior vena cava with a frequency gradient being present between the arrhythmogenic thoracic vein and the atria (Figure 1.2) [28].

More insight into underlying mechanisms came from a recent biatrial frequency mapping study combined with pharmacological intervention [30]. In this study, in patients with paroxysmal AF, adenosine infusion increased local dominant frequency, particularly at the PV left atrial junction, amplifying a left-to-right frequency gradient. In patients with permanent AF, dominant frequencies were significantly higher than in patients with paroxysmal AF in all atrial regions analyzed, with the highest adenosine increase of frequencies outside the PV region. Results of this study strongly suggest that AF is maintained by reentrant sources, most likely located at the PV left atrial junction in paroxysmal AF, whereas non-PV locations are more likely drivers in persistent AF.

From an electrophysiological point of view it can therefore be concluded that AF is not a homogeneous arrhythmia. This becomes further obvious by the different appearances of the fibrillatory waveforms on the surface ECG, ranging from no or very small waves ("fine AF") to large waves ("coarse AF"). In addition, high-density body surface mapping reveals different activation patterns being in close agreement with invasive mapping (Figure 1.3) [31].

Unfortunately, this heterogeneity is not reflected in the choice of therapy which may be considered as being based on trial and error. Hopefully, a better understanding of AF mechanisms and subsequently diagnostic tools to characterize them will become available to guide AF management [32].

1.3   CLASSIFICATION SCHEMES

There is no common agreement today on the best AF classification, although several schemes have been proposed in recent years. However, none of those succeeded to completely include the many different aspects of AF [33, 34, 35, 36]. Most common classification schemes are based on the electrocardiographic onset [36, 37] or on epicardial and endocardial electrograms obtained with different mapping systems [23, 38]. Gallagher and Camm divided AF into [36]:

1. paroxysmal AF, i.e., AF with spontaneous interruption generally within 7 days but mostly in 24–48 h,

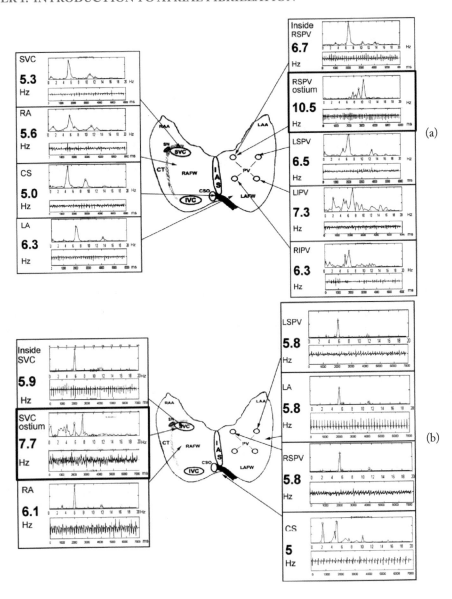

Figure 1.2: Paroxysmal AF originating from (a) the right superior pulmonary vein (RSPV) ostium and (b) ostium superior vena cava (SVC). Note the different frequency gradients depending on the driving source. Bipolar electrograms are recorded at different atrial and venous sites depicted schematically in the middle panel. Those are then subjected to Fourier analysis and the dominant frequencies are determined from the corresponding power spectra.

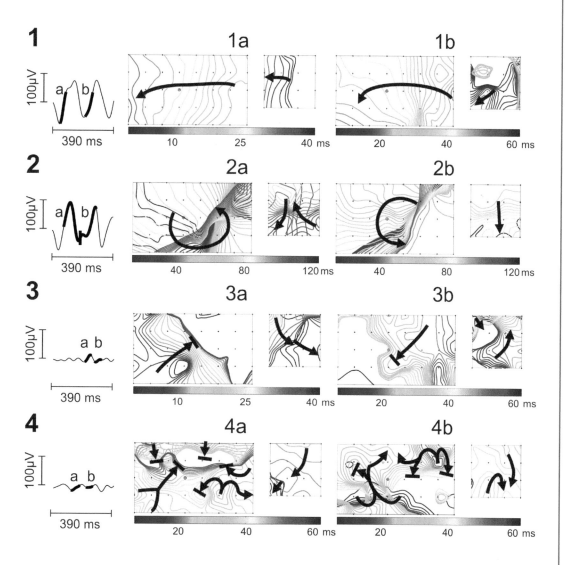

Figure 1.3: High-density activation mapping reveals different AF patterns from the surface ECG. As with invasive mapping, there are different degrees of activation complexity [31]. Left: Four examples of fibrillatory waves in TQ segments (lead V1). Right: Panels a and b correspond to wavefront propagation maps of two fibrillatory intervals (thick lines in the ECG). Maps are reconstructed from 40 chest leads positioned in the front (larger box), and 20 chest leads in the back of the thorax (smaller box). Arrows indicate the direction of propagation of each wavefront.

2. persistent AF, i.e., AF that, independently of its duration, does not interrupt spontaneously but with therapeutical interventions (pharmacological or electrical), and

3. permanent or chronic AF, i.e., AF in which interruption attempts have not been made or, if made, they were not successful, i.e., failed restoration of sinus rhythm or AF recidivism, or in which further cardioversion attempts are not warranted.

Moreover, the AF guidelines published in 2006 distinguished AF in [37]:

1. first detected AF episode, which could be paroxysmal, persistent, or permanent, and

2. recurring AF in the presence of two or more relapses.

These systematic efforts have had not only the important value to conform the 'semantics,' but also to allow a better comparison of published studies. It needs to be emphasized, however, that particular patients may have AF episodes that fall into different categories and that AF patterns may change over time.

1.4   EPIDEMIOLOGY

It has been estimated that the prevalence of AF in the US is about 2.2 million including paroxysmal or persistent AF (0.89% of the population in 1997) [39]. Similarly, the ATRIA study [40] observed a prevalence of AF of 0.95% in the North American general population; congruent data have also been obtained in the UK [41]. There are about 160,000 new AF cases each year only in the US and similar figures are also present in European countries.

From the Framingham experience [42, 43], we have learned that the prevalence of AF increases with aging of the population and doubles for every increase of decade after the 50 years in the population examined, reaching almost 10% in octogenarians. Those findings were also present in other epidemiological studies, such as the ATRIA study [40] where the AF prevalence was 0.1% in subjects younger than 55 years of age and 9% in patients older than 80 years.

A recent prospective study from the Netherlands, based on a cohort of subjects older than 55 years of age, has highlighted a total prevalence of AF of 5.5%, increasing from 0.7% in subjects from 55–59 years to be as high as 17.8% in 85 year old patients (Figure 1.4) [44].

It has been observed that AF is more frequent in men than in women (ratio 1.7) [44] and, in particular, in a trial on patients over 65 years AF prevalence was 9.1% in men and 4.8% in women [45]. The rationale behind the described greater susceptibility to AF onset in men than in women is currently unknown.

The prevalence of AF has progressively increased in recent years [46] and this trend will probably continue in the future, mainly because of the progressive fast aging of our population. It is projected that the number of affected individuals in the US will increase from 2.3 to 12.1–15.9 million by the year 2050 [47], thus determining a kind of a new epidemic [48].

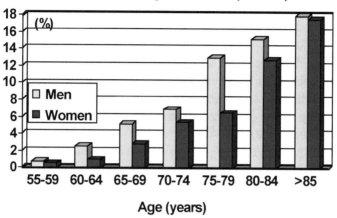

Prevalence of Atrial Fibrillation
The Rotterdam Study 1990 - 1993 (N=6808)

Figure 1.4: Prevalence of atrial fibrillation. (Adapted from [44].)

In the Framingham study, the incidence of AF was about 2% for both paroxysmal and persistent AF over a 20 year follow-up period [49]. This cumulative incidence corresponds to a punctual one of about 0.2% per year for both AF forms together.

As with AF prevalence, its incidence increases with the progressive aging of the population and doubles for every decade after the 50 years. From available data [42, 50], we can expect 10% of subjects over 80 years having at least one AF episode. On this basis, people older than 40 years have a 25% risk for developing at least one episode of AF. This risk remains globally high (16.6%) even when predisposing factors are statistically evaluated [51].

1.5 CLINICAL PRESENTATION

The typical patient with AF is often referred to as an elder one with diabetes, left ventricular hypertrophy (LVH), and/or other electrocardiographic pathological findings, coronary heart disease (CHD) or valvular heart disease, coronary heart failure (CHF), or a history of previous stroke [52].

The predisposing risk factors for AF onset are age, diabetes, electrocardiographic findings of LVH, arterial hypertension, body mass index, and, in women, cigarette smoking [50, 53, 54]. The Cardiovascular Health Study showed that 94% of women and 91% of men with AF had also a clinical or subclinical cardiac disease such as hypertension, valvular heart disease, coronary artery disease, dilatative or hypertrophic cardiomyopathy, while in the remaining no overt heart disease is present and AF is classified as "lone" [45].

In clinical trials, lone AF represents 15% of persistent AF and near to 60% of paroxysmal AF [55], while in population studies the prevalence of lone AF is lower (7.6–11% of persistent AF and 2.6% of paroxysmal AF in subjects younger than 60 years of age) [45, 56, 57].

Atrial fibrillation is frequently associated with a diversity of symptoms. In the ALFA study [58], palpitations were present in 54% of patients with AF, dyspnoea in 44.4%, exercise intolerance and asthenia in 14.3%, while angina (10.1%) and syncope (10.4%) were less frequently associated with AF. Palpitations were symptoms more frequently described in paroxysmal and new onset AF (79% and 51.5%, respectively).

In fact, the AF incidence and prevalence may be underestimated because of the complete lack of symptoms in some patients or because of the possibility of asymptomatic episodes in subjects otherwise symptomatic. In the Cardiovascular Heart Study [45], 30% of enrolled patients had an AF episode discovered by chance during an electrocardiographic control for other clinical reasons. These data were confirmed by the SPAF–III trial where this finding was observed in 45% [59].

In the ALFA study [58], 11% of patients with AF were asymptomatic. The Euro Heart Survey Trial on Atrial Fibrillation showed that more than the 70% of patients with paroxysmal, persistent, or new onset AF were symptomatic, while only 55% of patients with permanent AF had symptoms [54]. Moreover, 16% of patients with newly discovered AF, 6% with paroxysmal AF, 10% with persistent AF, and 21% with permanent AF were completely asymptomatic. In another study using frequent Holter monitoring during follow-up, the ratio of symptomatic and asymptomatic AF episodes was 1:12 [60].

This finding has important therapeutic implications, especially for initiation or possible discontinuation of antithrombotic therapy which apparently can not be based on symptoms. Progress in transtelephonic and telemetry monitoring techniques will certainly make a more realistic recognition of asymptomatic AF burden possible [61]. Several studies evaluated the effectiveness of antiarrhythmic or nonpharmacological treatments using ambulatory or transtelephonic recordings as indispensable tools to reach studies' endpoints [62, 63, 64, 65, 66]. These trials showed that about 17% of symptomatic AF recurrences preceded corresponding asymptomatic and that about 70% of the AF recurrences were identifiable only with these methods.

Further information has been obtained by analyzing data stored in pacemaker memory. Asymptomatic recurrences of AF, sometimes also of prolonged duration (>48 hours), have been observed in about a third of these subjects [67].

The quality of life in subjects with AF is markedly reduced if compared with controls, with a score index decrease of −16 to −30% when considering all usual and common parameters (i.e., general state of health, physical functions, vitality, mental state, emotional functions, social role, physical pain). Worsening of quality of life scores of patients with AF was more pronounced than that of patients with coronary heart disease, heart failure, or undergoing percutaneous coronary interventions [68].

1.6   MORTALITY AND MORBIDITY

Atrial fibrillation is not only related with frequent symptoms but also increases risk of death and stroke. In the Framingham trial, an increase in global and cardiovascular mortality was observed in low-risk patients with 'lone' AF when compared with control subjects without this arrhythmia [69]. In subjects from 55 to 94 years, AF represents an independent risk factor for the risk of death.

A recent case-control study was conducted on 577 patients with AF or atrial flutter and patients were followed for a mean period of 3.6±2.3 years [70]. This trial showed an increased risk of global mortality in AF patients of 7.8 times after 6 months and of 2.5 times at the end of the follow-up period, which was still present after correction for main known cardiovascular risk factors (95% confidence interval: 1.9–3.3). Similarly, a prospective trial, conducted on 15,000 Scotsmen for over 20 years showed a relative risk of all-cause mortality in patients with AF of 1.5 in men and of 2.2 in women confirming the Framingham data [71].

Mortality was particularly high in patients with heart disease, above all in subjects with CHF [69]. The new onset of CHF in a patient with AF determines an increase in the risk of death of 2.7% in men and of 3.1% in women. Analogously, the onset of a new episode of AF in patients with CHF determines an increase in the risk of death of 1.6% in men and of 2.7% in women. On the contrary, prognosis of lone AF appears more favorable with a risk of death similar to subjects without AF [41].

In recent years, equivalence between rhythm and rate control in patients over 65 years has been proposed [72, 73, 74, 75, 76]. Even though this observation was based on the probability of cardiovascular events, epidemiological data seem not to confirm the described benignity of this cardiac arrhythmia. In this context, it has to be mentioned that the authors of the AFFIRM trial performed a subanalysis of their study, showing an important reduction in the risk of death in patients remaining in sinus rhythm [77].

Atrial fibrillation is an independent risk factor for stroke. It is associated with a four- to fivefold increased risk of cerebrovascular events [43, 78, 79, 80], being responsible for a total of 15–18% of all strokes [79, 81]. The annual rate of thromboembolic events is considerably higher in patients with AF than in control subjects (4.5% versus 0.2–1.4%), with an invalidating stroke incidence of 2.5% [82, 83]. Considering transient ischemic attacks and silent strokes, this percentage increases to about 7% [83]. Patients with paroxysmal AF have an annual frequency of stroke (3.2%) comparable to that of patients with chronic or persistent forms (3.3%) [84].

The risk of stroke increases with age: elderly patients are more prone not only to develop AF, but also their risk of stroke is considerably higher than in younger subjects with AF [78]. The predisposition of elderly patients to develop an embolic stroke may also depend on higher prevalence of co-morbidities as well as on other risk factors such as valvular heart disease, arterial hypertension, CHF, previous cerebrovascular accidents, and diabetes mellitus [85].

1.7   TREATMENT OF ATRIAL FIBRILLATION

In the following section, a brief overview of general AF treatment options is given, followed by the introduction of an innovative treatment concept based on signal-guided therapies derived from the ECG and the intra-atrial electrogram.

GENERAL TREATMENT OPTIONS

Treatment of AF has three major objectives: prevention of thromboembolic complications, control of ventricular rate, and/or restoration of sinus rhythm.

Prophylaxis of Thromboembolic Complications

How to perform, when to start, and when to interrupt drug prophylaxis for preventing thromboembolic events are most critical decisions to make in all patients with AF. Regarding this point, it is worth to recall that the major difference in the outcome of patients enrolled in the AFFIRM study [77] was related to more efficient use of oral anticoagulant drugs in the rate control in comparison to rhythm control.

The thromboembolic risk increases significantly when AF duration exceeds 48 h. Thus, patient's management varies in relation to arrhythmia duration and attempts to restore sinus rhythm should be completed, when possible, within this time frame. An emerging concept is that the great incidence of asymptomatic episodes in paroxysmal AF or of asymptomatic recurrences after DC cardioversion cast doubts on the decision of the best timing for interrupting oral anticoagulant therapy in this group of patients, and also in patients who underwent radiofrequency ablation. Most patients with permanent AF have to continue oral anticoagulant therapy for the rest of their life.

Pharmacological Rate Control

Atrial fibrillation is generally characterized by a faster ventricular response than during sinus rhythm, particularly in young subjects. Thus, control of ventricular response is one of the initial objectives of pharmacological management of AF. In patients with permanent AF, pharmacological attempts to control and regularize ventricular response are associated with symptom decrease and better hemodynamic function. Drugs acting on the atrioventricular (AV) node are used in controlling ventricular response. Among them, digoxin, beta-blockers, and calcium antagonist are the most commonly applied [86], but sometimes amiodarone or AV node ablation is indicated in refractory cases.

Restoration of Sinus Rhythm

Sinus rhythm can be restored by means of two principal modalities: pharmacological and electrical cardioversion. The first one is generally performed by using class IC or III antiarrhythmic drugs. Both oral and intravenous administration routes have been proven effective; the latter one is generally associated with an early response. DC cardioversion is generally performed in patients with AF duration greater than 48 h under deep sedation requiring a short hospital stay. Biphasic shock is 98% effective in restoring sinus rhythm.

Pharmacological Prophylaxis of AF Recurrences

In about 60% of patients in whom sinus rhythm is restored with pharmacological or electrical cardioversion, AF recurs within 6–12 months [87]. Prophylaxis of AF recurrences is a major pharmacological target which, however, suffers from limited efficacy and relevant incidence of side effects of antiarrhythmic drugs. In addition to traditional class IC and III antiarrhythmic drugs [88, 89], other pharmacological interventions have been proposed as novel strategies to limit arrhythmia recurrences.

Among them, blockade of the renin–angiotensin–aldosterone system [90, 91] and statins [92] are the most promising, but clinical evaluation in large controlled trials is still missing.

Radiofrequency Catheter Ablation
After the original description of Haissaguerre et al. pointing to the proarrhythmic role of PV foci in inducing and maintaining AF, catheter ablation of AF triggers and substrates has become one of the most promising therapeutic approaches to AF management [93]. In this context, it is important to affirm that appropriate indications for this emerging treatment have been published [94]. Since its introduction in 1997, multiple approaches have been developed with similar success rates exceeding 80% in patients with paroxysmal AF and 50% in persistent AF [95].

New ablation techniques are currently under intense experimental and clinical investigation, mainly including alternative energy sources applied with balloon-based ablation systems and cardiac image integration.

SIGNAL-GUIDED THERAPY

As illustrated by other contributions in this book, numerous signal processing techniques exist for analysis and characterization of atrial activity and ventricular response in AF using different acquisition techniques. Interestingly, the mechanisms behind the various electrogram and ECG appearances, and their possible prognostic information contained therein, have just begun to be explored. In fact, a limited number of clinical studies have suggested electrogram and ECG parameters for clinical decision-making that will be summarized below.

It has been hypothesized that identification of ECG patterns of atrial activation during AF might become of critical importance in targeting those areas responsible for the maintenance of reentry circuits (as opposed to the identification of AF initiating triggers) that could be isolated with radiofrequency ablation. In fact, Nademanee et al. described an ablation approach targeting complex fragmented atrial electrograms which they identified by three-dimensional electroanatomical mapping (CARTO; Biosense Webster) in 121 patients with paroxysmal ($n = 57$) or persistent ($n = 64$) AF [96]. The selection of this target was based on previous observations from fragmented electrograms (Figure 1.5), defined as multicomponent atrial electrograms including [97]:

1. atrial electrograms with two or more deflections and/or perturbation of the baseline and/or continuous electrical activity over a 10-s period, or

2. atrial electrograms with very short cycle length ($\leq$120 ms) over a 10-s period reflect areas of slow conduction and/or pivot points indicating important areas for sustenance of reentry.

During AF, fragmented electrograms were predominantly found in the interatrial septum, PVs, left atrial roof, left posteroseptal mitral annulus, and coronary sinus ostium. Their elimination by radiofrequency ablation terminated AF in 95%, and resulted in freedom from AF and symptoms in 91%.

Similarly, Sanders et al. studied sites of dominant activation frequency and the effect of ablation at these sites [98]. In accordance with the aforementioned frequency mapping studies, the authors

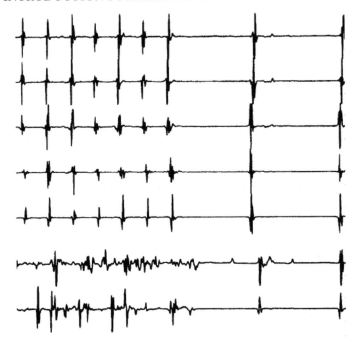

Figure 1.5: Ablation of complex fractionated 2-s electrograms as an example of signal-guided AF therapy. Multiple bipolar electrograms (top 5 tracings) from the right atrium show a very organized (type I) fibrillatory activity. In contrast, complex fractionated electrograms are found in the left superior pulmonary vein (bottom 2 tracings). Radiofrequency ablation at this site terminates AF as reflected by the last two beats.

noted different frequency distributions in paroxysmal and persistent AF. Interestingly, ablation at a site with the dominating fibrillatory frequency resulted in significant prolongation of the AF cycle length and AF terminated subsequently in 17 of 19 patients with paroxysmal, but in 0 of 13 with persistent AF.

Besides being only moderately effective, the utilization of antiarrhythmic drugs may be associated with serious side effects including life-threatening ventricular arrhythmias and severe extra-cardiac side effects requiring their discontinuation. For this and other reasons, electrocardiographic approaches that might help differentiating responders from nonresponders to antiarrhythmic drug therapy, or selecting patients undergoing cardioversion who may also benefit from long-term drug treatment, are of special interest and clinical need.

At the atrial level, characterization of fibrillatory waves during AF using spectral analysis or P wave signal averaging during (restored) sinus rhythm have been successfully applied to monitor and predict outcome of antiarrhythmic drug therapy and cardioversion.

A substantial reduction in atrial fibrillatory rates following different intravenously or orally administered class I and III antiarrhythmic drugs as well as following verapamil or magnesium has been reported using serial or continuous spectral analysis of the surface ECG [99]. Time–frequency analysis can be used to quantify instantaneous changes in fibrillatory rate that may be associated with intervention such as monitoring antiarrhythmic drug effects during their loading phase [100, 101]; see Chapter 4.

Moreover, a baseline rate of 360 fibrillations per minute was predictive of AF termination following intravenous ibutilide [102] or oral flecainide [103]. Other authors noted no baseline fibrillatory rate difference between patients who converted to sinus rhythm and those who did not after oral bepridil [104] or intravenous ibutilide [105] administration. Instead, larger and more rapid rate decreases were associated with AF termination.

The outcome of cardioversion in terms of atrial defibrillation threshold and AF recurrence in relation to baseline fibrillatory rate has also been reported. In particular, higher fibrillatory rates were observed in patients with early arrhythmia recurrence immediately prior to internal [106] or external cardioversion [107, 108, 109] when compared with patients who maintained sinus rhythm.

Several studies have attempted to define the role of P wave signal averaging in monitoring and predicting antiarrhythmic drug efficacy [110, 111, 112, 113] and AF recurrence following cardioversion [114, 115, 116, 117, 118]. For example, after initiation of disopyramide in patients with paroxysmal AF, filtered P wave duration (PWD) was prolonged and remained prolonged after 4 weeks of treatment [110], while PWD and atrial late potentials remained unchanged. In another study, atrial late potentials decreased after amiodarone [113]. Of special interest is the different behavior of responders and nonresponders. Increased PWD dispersion after disopyramide [110] or pilsicainide [111] was strongly associated with AF recurrence, as opposed to decreased PWD dispersion in patients who remained in sinus rhythm. Similarly, decreased PWD and reduced late potentials after amiodarone loading were associated with fewer AF relapses [113].

At the ventricular level, analysis of heart rate variability (HRV) has been applied to predict AF onset and recurrence. As with P wave signal averaging, the majority of recordings were made during sinus rhythm in paroxysmal or after cardioversion in persistent AF.

The prognostic value of HRV for predicting arrhythmia recurrence after successful cardioversion has been analyzed. Signs of increased sympathetic and reduced vagal modulation (LF/HF ratio >2) at the time of discharge after successful cardioversion was associated with early but not late AF recurrence [119]. In contrast, persistently higher values of 24-h time domain parameters reflecting vagal modulation have been reported in patients with AF recurrences after internal cardioversion performed under general anaesthesia [120]. Transient alterations of autonomic modulation rather than a specific autonomic pattern seem to favor both arrhythmia onset and recurrence in most patients.

In Poincaré plots, each RR interval is plotted against the preceding RR interval; see Chapter 5. By including all RR interval pairs, a plot can be constructed in which clusters of RR intervals can be identified. Interestingly, in AF patients with clustering of RR intervals, electrical cardioversion was more effective to restore sinus rhythms, and, of greater clinical interest, sinus rhythm persisted for a longer period than in patients without clustering [121].

Finally, characterization of ventricular response using RR interval histogram analysis has been used to predict outcome of AV nodal modification for ventricular rate control and mortality. When constructing RR interval histograms obtained from Holter recordings with persistent AF, uni-, bi-, or multimodal RR distribution patterns can be found [122, 123]. In up to 55% of patients, the ventricular response during AF evidences two separate populations of RR intervals already at a naked eye inspection of the ECG [124]. The two different RR populations are suggested to correspond to conduction along two different atrionodal conduction routes [122], although recent work from our group suggests that preferential AV conduction ratios may be the underlying mechanism [125]. Nevertheless, some investigators analyzed the correlation between RR interval histogram pattern and efficacy of radiofrequency modification of AV node in patients with chronic AF [124, 126]. These authors observed that radiofrequency modification of AV node was more effective in controlling ventricular response in patients with bimodal distributions of RR interval pattern. Interference with posterior atrionodal input is considered to be the prevailing mechanism of rate control in these patients.

Reduced ventricular response irregularity has been associated with increased mortality in patients with chronic AF. Both variability and irregularity parameters were significant predictors of mortality at univariate analysis. After adjustment for clinical variables, however, only irregularity indices such as approximate entropy maintained their independent prognostic value for cardiac mortality [127].

1.8   CONCLUSIONS

The importance of AF has been largely underestimated for several decades and, probably, this is the moment of reconsidering with great attention its epidemiological profile.

There is nowadays growing attention for the costs of AF management and of its possible complications, in particular stroke and heart failure. As a fact, AF is the most common sustained cardiac arrhythmia in daily clinical practice and it may determine several important complications. A detailed understanding of the epidemiology of AF is essential for planning adequate diagnostic and therapeutical pathways and consequently for cost saving. In recent years, several important changes in AF prevalence and incidence have been observed. Confident projections forecast an epidemical explosion of new cases of AF soon, mainly because of progressive aging of our populations.

Atrial fibrillation is associated with a variety of clinical manifestations and different treatment options are available. Development of new and easy to use algorithms for classification of fibrillatory waves and ventricular response patterns are examples of possible collaboration between clinicians and engineers. Hopefully, in the near future, bedside examination of any AF patient will include direct information on individual electrophysiological, structural, and autonomic characteristics which, at the moment, can only be obtained in a limited number of patients.

Bibliography

[1] G. Boriani, I. Diemberger, C. Martignani, M. Biffi, and A. Branzi, "The epidemiological burden of atrial fibrillation: A challenge for clinicians and health care systems," *Eur. Heart J.*, vol. 27, pp. 893–894, 2006. DOI: 10.1093/eurheartj/ehi651

[2] N. Wiener and A. Rosenblueth, "The mathematical formulation of the problem of conduction of impulses in a network of connected excitable elements specifically in cardiac muscle," *Arch. Inst. Cardiol. Mex.*, vol. 16, pp. 205–265, 1946.

[3] C. A. Morillo, G. J. Klein, D. L. Jones, and C. M. Guiraudon, "Chronic rapid atrial pacing. Structural, functional, and electrophysiological characteristics of a new model of sustained atrial fibrillation," *Circulation*, vol. 91, pp. 1588–1595, 1995.

[4] M. C. Wijffels, C. J. Kirchhof, R. Dorland, and M. A. Allessie, "Atrial fibrillation begets atrial fibrillation. A study in awake chronically instrumented goats," *Circulation*, vol. 92, pp. 1954–1968, 1995.

[5] M. C. Wijffels, C. J. Kirchhof, R. Dorland, J. Power, and M. A. Allessie, "Electrical remodeling due to atrial fibrillation in chronically instrumented conscious goats: Roles of neurohumoral changes, ischemia, atrial stretch, and high rate of electrical activation," *Circulation*, vol. 96, pp. 3710–3720, 1997.

[6] A. Elvan, K. Wylie, and D. P. Zipes, "Pacing-induced chronic atrial fibrillation impairs sinus node function in dogs. Electrophysiological remodeling," *Circulation*, vol. 94, pp. 2953–2960, 1996.

[7] R. Gaspo, R. F. Bosch, M. Talajic, and S. Nattel, "Functional mechanisms underlying tachycardia-induced sustained atrial fibrillation in a chronic dog model," *Circulation*, vol. 96, pp. 4027–4035, 1997.

[8] E. G. Daoud, F. Bogun, R. Goyal, M. Harvey, K. C. Man, S. A. Strickberger, and F. Morady, "Effect of atrial fibrillation on atrial refractoriness in humans," *Circulation*, vol. 94, pp. 1600–1606, 1996.

[9] E. G. Daoud, B. P. Knight, R. Weiss, M. Bahu, W. Paladino, R. Goyal, C. Man, S. A. Strickberger, and F. Morady, "Effect of verapamil and procainamide on atrial fibrillation-induced electrical remodeling in humans," *Circulation*, vol. 96, pp. 1542–1550, 1997.

[10] W. C. Yu, S. A. Chen, S. H. Lee, C. T. Tai, A. N. Feng, B. I. Kuo, Y. A. Ding, and M. S. Chang, "Tachycardia-induced change of atrial refractory period in humans: Rate dependency and effects of antiarrhythmic drugs," *Circulation*, vol. 97, pp. 2331–2337, 1998.

[11] K. Kumagai, S. Akimitsu, K. Kawahira, F. Kawanami, Y. Yamanouchi, T. Hiroki, and K. Arakawa, "Electrophysiological properties in chronic lone atrial fibrillation," *Circulation*, vol. 84, pp. 1662–1668, 1991.

[12] M. R. Franz, P. L. Karasik, C. Li, J. Moubarak, and M. Chavez, "Electrical remodeling of the human atrium: Similar effects in patients with chronic atrial fibrillation and atrial flutter," *J. Am. Coll. Cardiol.*, vol. 30, pp. 1785–1792, 1997. DOI: 10.1016/S0735-1097(97)00385-9

[13] K. Kamalvand, K. Tan, G. Lloyd, J. Gill, C. Bucknall, and N. Sulke, "Alterations in atrial electrophysiology associated with chronic atrial fibrillation in man," *Eur. Heart J.*, vol. 20, pp. 888–895, 1999. DOI: 10.1053/euhj.1998.1404

[14] W. C. Yu, S. H. Lee, C. T. Tai, C. F. Tsai, M. H. Hsieh, C. C. Chen, Y. A. Ding, M. S. Chang, and S. A. Chen, "Reversal of atrial electrical remodeling following cardioversion of long-standing atrial fibrillation in man," *Cardiovasc. Res.*, vol. 42, pp. 470–476, 1999. DOI: 10.1016/S0008-6363(99)00030-9

[15] G. Boriani, M. Biffi, R. Zannoli, A. Branzi, and B. Magnani, "Evaluation of atrial refractoriness and atrial fibrillation inducibility immediately after internal cardioversion in patients with chronic persistent atrial fibrillation," *Cardiovasc. Drugs Ther.*, vol. 13, pp. 507–511, 1999. DOI: 10.1023/A:1007823619899

[16] B. J. Brundel, I. C. Van Gelder, R. H. Henning, R. G. Tieleman, A. E. Tuinenburg, M. Wietses, J. G. Grandjean, W. H. Van Gilst, and H. J. Crijns, "Ion channel remodeling is related to intraoperative atrial effective refractory periods in patients with paroxysmal and persistent atrial fibrillation," *Circulation*, vol. 103, pp. 684–690, 2001.

[17] E. G. Manios, E. M. Kanoupakis, G. I. Chlouverakis, M. D. Kaleboubas, H. E. Mavrakis, and P. E. Vardas, "Changes in atrial electrical properties following cardioversion of chronic atrial fibrillation: Relation with recurrence," *Cardiovasc. Res.*, vol. 47, pp. 244–253, 2000. DOI: 10.1016/S0008-6363(00)00100-0

[18] Y. Tanabe, M. Chinushi, K. Taneda, S. Fujita, H. Kasai, M. Yamaura, S. Imai, and Y. Aizawa, "Recovery of the right atrial effective refractory period after cardioversion of chronic atrial fibrillation," *Am. J. Cardiol.*, vol. 84, pp. 1261–1264, 1999. DOI: 10.1016/S0002-9149(99)00544-5

[19] M. Nishino, S. Hoshida, J. Tanouchi, T. Ito, J. Kato, K. Iwai, H. Tanahashi, M. Hori, Y. Yamada, and T. Kamada, "Time to recover from atrial hormonal, mechanical, and electrical dysfunction

after successful electrical cardioversion of persistent atrial fibrillation," *Am. J. Cardiol.*, vol. 85, pp. 1451–1454, 2000. DOI: 10.1016/S0002-9149(00)00793-1

[20] R. G. Tieleman, I. C. van Gelder, H. J. Crijns, P. J. de Kam, M. P. van den Berg, J. Haaksma, J. J. van der Woude, and M. A. Allessie, "Early recurrences of atrial fibrillation after electrical cardioversion: A result of fibrillation-induced electrical remodeling of the atria?," *J. Am. Coll. Cardiol.*, vol. 31, pp. 167–173, 1998. DOI: 10.1016/S0735-1097(97)00455-5

[21] G. Moe, "On the multiple wavelet hypothesis of atrial fibrillation," *Arch. Int. Pharmacodyn. Ther.*, vol. 140, pp. 183–188, 1962.

[22] M. A. Allessie, W. J. Lammers, F. I. Bonke, and J. M. Hollen, "Experimental evaluation of Moe's multiple wavelet hypothesis of atrial fibrillation," in *Cardiac Electrophysiology and Arrhythmias* (D. P. Zipes and J. Jaliff, eds.), pp. 265–275, Grune & Stratton, Orlando, Florida, 1985.

[23] K. T. Konings, C. J. Kirchhof, J. R. Smeets, H. J. Wellens, O. C. Penn, and M. A. Allessie, "High-density mapping of electrically induced atrial fibrillation in humans," *Circulation*, vol. 89, pp. 1665–1680, 1994.

[24] M. Holm, R. Johansson, J. Brandt, C. Lührs, and S. B. Olsson, "Epicardial right atrial free wall mapping in chronic atrial fibrillation. Documentation of repetitive activation with a focal spread–a hitherto unrecognised phenomenon in man," *Eur. Heart J.*, vol. 18, pp. 290–310, 1997.

[25] T. Sueda, K. Imai, O. Ishii, K. Orihashi, M. Watari, and K. Okada, "Efficacy of pulmonary vein isolation for the elimination of chronic atrial fibrillation in cardiac valvular surgery," *Ann. Thorac. Surg.*, vol. 71, pp. 1189–1193, 2001. DOI: 10.1016/S0003-4975(00)02606-0

[26] T. J. Wu, R. N. Doshi, H. L. Huang, C. Blanche, R. M. Kass, A. Trento, W. Cheng, H. S. Karagueuzian, C. T. Peter, and P. S. Chen, "Simultaneous biatrial computerized mapping during permanent atrial fibrillation in patients with organic heart disease," *J. Cardiovasc. Electrophysiol.*, vol. 13, pp. 571–577, 2002. DOI: 10.1046/j.1540-8167.2002.00571.x

[27] J. Sahadevan, K. Ryu, L. Peltz, C. M. Khrestian, R. W. Stewart, A. H. Markowitz, and A. L. Waldo, "Epicardial mapping of chronic atrial fibrillation in patients: Preliminary observations," *Circulation*, vol. 110, pp. 3293–3299, 2004. DOI: 10.1161/01.CIR.0000147781.02738.13

[28] Y. J. Lin, C. T. Tai, T. Kao, H. W. Tso, S. Higa, H. M. Tsao, S. L. Chang, M. H. Hsieh, and S. A. Chen, "Frequency analysis in different types of paroxysmal atrial fibrillation," *J. Am. Coll. Cardiol.*, vol. 47, pp. 1401–1407, 2006. DOI: 10.1016/j.jacc.2005.10.071

[29] S. Lazar, S. Dixit, F. E. Marchlinski, D. J. Callans, and E. P. Gerstenfeld, "Presence of left-to-right atrial frequency gradient in paroxysmal but not persistent atrial fibrillation in humans," *Circulation*, vol. 110, pp. 3181–3186, 2004. DOI: 10.1161/01.CIR.0000147279.91094.5E

[30] F. Atienza, J. Almendral, J. Moreno, R. Vaidyanathan, A. Talkachou, J. Kalifa, A. Arenal, J. P. Villacastin, E. G. Torrecilla, A. Sanchez, R. Ploutz-Snyder, J. Jalife, and O. Berenfeld, "Activation of inward rectifier potassium channels accelerates atrial fibrillation in humans: Evidence for a reentrant mechanism," *Circulation*, vol. 114, pp. 2434–2442, 2006. DOI: 10.1161/CIRCULATIONAHA.106.633735

[31] M. S. Guillem, A. M. Climent, F. Castells, D. Husser, J. Millet, and A. Bollmann, "Noninvasive, high-density mapping of human atrial fibrillation. Introduction and illustration of a novel diagnostic tool," in *Proc. Comput. Cardiol.*, vol. 34, http://cinc.mit.edu, 2007.

[32] A. Bollmann, "First comes diagnosis then comes treatment: An underappreciated paradigm in atrial fibrillation management," *Eur. Heart J.*, vol. 26, pp. 2487–2489, 2005. DOI: 10.1093/eurheartj/ehi578

[33] S. Levy, P. Novella, P. Ricard, and F. Paganelli, "Paroxysmal atrial fibrillation: A need for classification," *J. Cardiovasc. Electrophysiol.*, vol. 6, pp. 69–74, 1995. DOI: 10.1111/j.1540-8167.1995.tb00758.x

[34] S. M. Sopher and A. J. Camm, "Therapy for atrial fibrillation: Control of the ventricular response and prevention of recurrence," *Coron. Artery Dis.*, vol. 6, pp. 106–114, 1995. DOI: 10.1097/00019501-199502000-00004

[35] S. Levy, "Classification system of atrial fibrillation," *Curr. Opin. Cardiol.*, vol. 15, pp. 54–57, 2000. DOI: 10.1097/00001573-200001000-00007

[36] M. M. Gallagher and A. J. Camm, "Classification of atrial fibrillation," *Am. J. Cardiol.*, vol. 82, pp. 18N–28N, 1998. DOI: 10.1016/S0002-9149(98)00736-X

[37] V. Fuster, L. E. Ryden, D. S. Cannom, H. J. Crijns, A. B. Curtis, *et al.*, "ACC/AHA/ESC 2006 guidelines for the management of patients with atrial fibrillation: full text: A report of the American College of Cardiology/American Heart Association Task Force on practice guidelines and the European Society of Cardiology Committee for Practice Guidelines developed in collaboration with the European Heart Rhythm Association and the Heart Rhythm Society," *Europace*, vol. 8, pp. 651–745, 2006.

[38] J. L. Wells Jr., R. B. Karp, N. T. Kouchoukos, W. A. MacLean, T. N. James, and A. L. Waldo, "Characterization of atrial fibrillation in man: Studies following open heart surgery," *Pacing Clin. Electrophysiol.*, vol. 1, pp. 426–438, 1978. DOI: 10.1111/j.1540-8159.1978.tb03504.x

[39] S. S. Chugh, J. L. Blackshear, W. K. Shen, S. C. Hammill, and B. J. Gersh, "Epidemiology and natural history of atrial fibrillation: Clinical implications," *J. Am. Coll. Cardiol.*, vol. 37, pp. 371–378, 2001. DOI: 10.1016/S0735-1097(00)01107-4

[40] A. S. Go, E. M. Hylek, K. A. Phillips, Y. Chang, L. E. Henault, J. V. Selby, and D. E. Singer, "Prevalence of diagnosed atrial fibrillation in adults: National implications for rhythm management and stroke prevention: The AnTicoagulation and Risk Factors in Atrial Fibrillation (ATRIA) Study," *JAMA*, vol. 285, pp. 2370–2375, 2001. DOI: 10.1001/jama.285.18.2370

[41] S. Stewart, N. Murphy, A. Walker, A. McGuire, and J. J. V. McMurray, "Cost of an emerging epidemic: An economic analysis of atrial fibrillation in the UK," *Heart*, vol. 90, pp. 286–292, 2004. DOI: 10.1136/hrt.2002.008748

[42] W. B. Kannel, P. A. Wolf, E. J. Benjamin, and D. Levy, "Prevalence, incidence, prognosis, and predisposing conditions for atrial fibrillation: Population-based estimates," *Am. J. Cardiol.*, vol. 82, pp. 2N–9N, 1998. DOI: 10.1016/S0002-9149(98)00583-9

[43] P. A. Wolf, R. D. Abbott, and W. B. Kannel, "Atrial fibrillation as an independent risk factor for stroke: The Framingham Study," *Stroke*, vol. 22, pp. 983–988, 1991.

[44] J. Heeringa, D. A. van der Kuip, A. Hofman, J. A. Kors, G. van Herpen, B. H. Stricker, T. Stijnen, G. Y. Lip, and J. C. Witteman, "Prevalence, incidence and lifetime risk of atrial fibrillation: The Rotterdam study," *Eur. Heart J.*, vol. 27, pp. 949–953, 2006. DOI: 10.1093/eurheartj/ehi825

[45] C. D. Furberg, B. M. Psaty, T. A. Manolio, J. M. Gardin, V. E. Smith, and P. M. Rautaharju, "Prevalence of atrial fibrillation in elderly subjects (the Cardiovascular Health Study)," *Am. J. Cardiol*, vol. 74, pp. 236–241, 1994. DOI: 10.1016/0002-9149(94)90363-8

[46] P. A. Wolf, E. J. Benjamin, A. J. Belanger, W. B. Kannel, D. Levy, and R. B. D'Agostino, "Secular trends in the prevalence of atrial fibrillation: The Framingham Study," *Am. Heart J.*, vol. 131, pp. 790–795, 1996. DOI: 10.1016/S0002-8703(96)90288-4

[47] Y. Miyasaka, M. E. Barnes, B. J. Gersh, S. S. Cha, K. R. Bailey, W. P. Abhayaratna, J. B. Seward, and T. S. Tsang, "Secular trends in incidence of atrial fibrillation in Olmsted County, Minnesota, 1980 to 2000, and implications on the projections for future prevalence," *Circulation*, vol. 114, pp. 119–125, 2006. DOI: 10.1161/CIRCULATIONAHA.105.595140

[48] T. S. Tsang and B. J. Gersh, "Atrial fibrillation: An old disease, a new epidemic," *Am. J. Med.*, vol. 113, pp. 432–435, 2002. DOI: 10.1016/S0002-9343(02)01245-7

[49] W. Kannel, R. Abbot, D. Savage, and P. McNamara, "Coronary heart disease and atrial fibrillation: The Framingham study," *Am. Heart J.*, vol. 106, pp. 389–396, 1983. DOI: 10.1016/0002-8703(83)90208-9

[50] E. J. Benjamin, D. Levy, S. M. Vaziri, R. B. D'Agostino, A. J. Belanger, and P. A. Wolf, "Independent risk factors for atrial fibrillation in a population-based cohort. The Framingham Heart Study," *JAMA*, vol. 271, pp. 840–844, 1994. DOI: 10.1001/jama.271.11.840

[51] D. M. Lloyd-Jones, T. J. Wang, E. P. Leip, M. G. Larson, D. Levy, R. S. Vasan, R. B. D'Agostino, J. M. Massaro, A. Beiser, P. A. Wolf, and E. J. Benjamin, "Lifetime risk for development of atrial fibrillation: The Framingham Heart Study," *Circulation*, vol. 110, pp. 1042–1046, 2004. DOI: 10.1161/01.CIR.0000140263.20897.42

[52] W. M. Feinberg, J. L. Blackshear, A. Laupacis, R. Kronmal, and R. G. Hart, "Prevalence, age distribution, and gender of patients with atrial fibrillation. Analysis and implications," *Arch. Intern. Med.*, vol. 155, pp. 469–473, 1995. DOI: 10.1001/archinte.155.5.469

[53] S. M. Vaziri, M. G. Larson, E. J. Benjamin, and D. Levy, "Echocardiographic predictors of nonrheumatic atrial fibrillation. The Framingham Heart Study," *Circulation*, vol. 89, pp. 724–730, 1994.

[54] R. Nieuwlaat, A. Capucci, A. J. Camm, S. B. Olsson, D. Andresen, D. W. Davies, S. Cobbe, G. Breithardt, J. Y. Le Heuzey, M. H. Prins, S. Levy, and H. J. Crijns, "Atrial fibrillation management: A prospective survey in ESC member countries: The Euro Heart Survey on Atrial Fibrillation," *Eur. Heart J.*, vol. 26, pp. 2422–2434, 2005. DOI: 10.1093/eurheartj/ehi505

[55] F. D. Murgatroyd and A. J. Camm, "Atrial arrhythmias," *Lancet*, vol. 341, pp. 1317–1322, 1993. DOI: 10.1016/0140-6736(93)90824-Z

[56] F. N. Brand, R. D. Abbott, W. B. Kannel, and P. A. Wolf, "Characteristics and prognosis of lone atrial fibrillation. 30-year follow-up in the Framingham Study," *JAMA*, vol. 254, pp. 3449–3453, 1985. DOI: 10.1001/jama.254.24.3449

[57] S. L. Kopecky, B. J. Gersh, M. D. McGoon, J. P. Whisnant, D. R. Holmes Jr., D. M. Ilstrup, and R. L. Frye, "The natural history of lone atrial fibrillation. A population-based study over three decades," *N. Engl. J. Med.*, vol. 317, pp. 669–674, 1987.

[58] S. Levy, M. Maarek, P. Coumel, L. Guize, J. Lekieffre, J. L. Medvedowsky, and A. Sebaoun, "Characterization of different subsets of atrial fibrillation in general practice in France: The ALFA study. The College of French Cardiologists," *Circulation*, vol. 99, pp. 3028–3035, 1999.

[59] J. L. Blackshear, S. L. Kopecky, S. C. Litin, R. E. Safford, and S. C. Hammill, "Management of atrial fibrillation in adults: Prevention of thromboembolism and symptomatic treatment," *Mayo Clin. Proc.*, vol. 71, pp. 150–160, 1996.

[60] R. L. Page, W. E. Wilkinson, W. K. Clair, E. A. McCarthy, and E. L. Pritchett, "Asymptomatic arrhythmias in patients with symptomatic paroxysmal atrial fibrillation and paroxysmal supraventricular tachycardia," *Circulation*, vol. 89, pp. 224–227, 1994.

[61] J. A. Olson, A. M. Fouts, B. J. Padanilam, and E. N. Prystowsky, "Utility of mobile cardiac outpatient telemetry for the diagnosis of palpitations, presyncope, syncope, and the assessment of therapy efficacy," *J. Cardiovasc. Electrophysiol.*, vol. 18, pp. 473–477, 2007. DOI: 10.1111/j.1540-8167.2007.00779.x

[62] J. L. Anderson, E. M. Gilbert, B. Alpert, R. W. Henthorn, A. L. Waldo, A. K. Bhandari, R. W. Hawkinson, and E. L. Pritchett, "Prevention of symptomatic recurrences of paroxysmal atrial fibrillation in patients initially tolerating antiarrhythmic therapy. A multicenter, double-blind, crossover study of flecainide and placebo with transtelephonic monitoring. Flecainide Supraventricular Tachycardia Study Group," *Circulation*, vol. 80, pp. 1557–1570, 1989.

[63] R. Wolk, P. Kulakowski, S. Karczmarewicz, G. Karpinski, E. Makowska, A. Czepiel, and L. Ceremuzynski, "The incidence of asymptomatic paroxysmal atrial fibrillation in patients treated with propranolol or propafenone," *Int. J. Cardiol.*, vol. 54, pp. 207–211, 1996. DOI: 10.1016/0167-5273(96)02631-9

[64] R. L. Page, T. W. Tilsch, S. J. Connolly, D. J. Schnell, S. R. Marcello, W. E. Wilkinson, and E. L. Pritchett, "Asymptomatic or 'silent' atrial fibrillation: Frequency in untreated patients and patients receiving azimilide," *Circulation*, vol. 107, pp. 1141–1145, 2003. DOI: 10.1161/01.CIR.0000051455.44919.73

[65] T. Fetsch, P. Bauer, R. Engberding, H. P. Koch, J. Lukl, T. Meinertz, M. Oeff, L. Seipel, H. J. Trappe, N. Treese, and G. Breithardt, "Prevention of atrial fibrillation after cardioversion: Results of the PAFAC trial," *Eur. Heart J.*, vol. 25, pp. 1385–1394, 2004. DOI: 10.1016/j.ehj.2004.04.015

[66] O. M. Wazni, N. F. Marrouche, D. O. Martin, A. Verma, M. Bhargava, W. Saliba, D. Bash, R. Schweikert, J. Brachmann, J. Günther, K. Gutleben, E. Pisano, D. Potenza, R. Fanelli, A. Raviele, S. Themistoclakis, A. Rossillo, A. Bonso, and A. Natale, "Radiofrequency ablation vs antiarrhythmic drugs as first-line treatment of symptomatic atrial fibrillation: A randomized trial," *JAMA*, vol. 293, pp. 2634–2640, 2005. DOI: 10.1001/jama.293.21.2634

[67] C. W. Israel, G. Gronefeld, J. R. Ehrlich, Y. G. Li, and S. H. Hohnloser, "Long-term risk of recurrent atrial fibrillation as documented by an implantable monitoring device: Implications for optimal patient care," *J. Am. Coll. Cardiol.*, vol. 43, pp. 47–52, 2004. DOI: 10.1016/j.jacc.2003.08.027

[68] P. Dorian, W. Jung, D. Newman, M. Paquette, K. Wood, G. M. Ayers, A. J. Camm, M. Akhtar, and B. Luderitz, "The impairment of health-related quality of life in patients with intermittent atrial fibrillation: Implications for the assessment of investigational therapy," *J. Am. Coll. Cardiol.*, vol. 36, pp. 1303–1309, 2000. DOI: 10.1016/S0735-1097(00)00886-X

[69] T. J. Wang, M. G. Larson, D. Levy, R. S. Vasan, E. P. Leip, P. A. Wolf, R. B. D'Agostino, J. M. Murabito, W. B. Kannel, and E. J. Benjamin, "Temporal relations of atrial fibrillation and congestive heart failure and their joint influence on mortality: The Framingham Heart Study," *Circulation*, vol. 107, pp. 2920–2925, 2003. DOI: 10.1161/01.CIR.0000072767.89944.6E

[70] H. Vidaillet, J. F. Granada, P. H. Chyou, K. Maassen, M. Ortiz, J. N. Pulido, P. Sharma, P. N. Smith, and J. Hayes, "A population-based study of mortality among patients with atrial fibrillation or flutter," *Am. J. Med.*, vol. 113, pp. 365–370, 2002. DOI: 10.1016/S0002-9343(02)01253-6

[71] S. Stewart, C. L. Hart, D. J. Hole, and J. J. McMurray, "A population-based study of the long-term risks associated with atrial fibrillation: 20-year follow-up of the Renfrew/Paisley study," *Am. J. Med.*, vol. 113, pp. 359–364, 2002. DOI: 10.1016/S0002-9343(02)01236-6

[72] S. H. Hohnloser, K. H. Kuck, and J. Lilienthal, "Rhythm or rate control in atrial fibrillation— Pharmacological Intervention in Atrial Fibrillation (PIAF): A randomised trial," *Lancet*, vol. 356, pp. 1789–1794, 2000. DOI: 10.1016/S0140-6736(00)03230-X

[73] J. Carlsson, S. Miketic, J. Windeler, A. Cuneo, S. Haun, S. Micus, S. Walter, and U. Tebbe, "Randomized trial of rate-control versus rhythm-control in persistent atrial fibrillation: The Strategies of Treatment of Atrial Fibrillation (STAF) study," *J. Am. Coll. Cardiol.*, vol. 41, pp. 1690–1696, 2003. DOI: 10.1016/S0735-1097(03)00332-2

[74] G. Opolski, A. Torbicki, D. A. Kosior, M. Szulc, B. Wozakowska-Kaplon, P. Kolodziej, and P. Achremczyk, "Rate control vs rhythm control in patients with nonvalvular persistent atrial fibrillation: The results of the Polish How to Treat Chronic Atrial Fibrillation (HOT CAFE) Study," *Chest*, vol. 126, pp. 476–486, 2004. DOI: 10.1378/chest.126.2.476

[75] D. G. Wyse, A. L. Waldo, J. P. DiMarco, M. J. Domanski, Y. Rosenberg, E. B. Schron, J. C. Kellen, H. L. Greene, M. C. Mickel, J. E. Dalquist, and S. D. Corley, "A comparison of rate control and rhythm control in patients with atrial fibrillation," *N. Engl. J. Med.*, vol. 347, pp. 1825–1833, 2002.

[76] I. C. van Gelder, V. E. Hagens, H. A. Bosker, J. H. Kingma, S. A. Said, J. I. Darmanata, A. J. Timmermans, J. G. Tijssen, and H. J. Crijns, "A comparison of rate control and rhythm control in patients with recurrent persistent atrial fibrillation," *N. Engl. J. Med.*, vol. 347, pp. 1834–1840, 2002. DOI: 10.1056/NEJMoa021375

[77] S. D. Corley, A. E. Epstein, J. P. DiMarco, M. J. Domanski, N. Geller, H. L. Greene, R. A. Josephson, J. C. Kellen, R. C. Klein, A. D. Krahn, M. Mickel, L. B. Mitchell, J. D. Nelson, Y. Rosenberg, E. Schron, L. Shemanski, A. L. Waldo, and D. G. Wyse, "Relationships between sinus rhythm, treatment, and survival in the Atrial Fibrillation Follow-Up Investigation of Rhythm Management (AFFIRM) Study," *Circulation*, vol. 109, pp. 1509–1513, 2004. DOI: 10.1161/01.CIR.0000121736.16643.11

[78] P. A. Wolf, T. R. Dawber, H. E. Thomas Jr., and W. B. Kannel, "Epidemiologic assessment of chronic atrial fibrillation and risk of stroke: The Framingham Study," *Neurology*, vol. 28, pp. 973–977, 1978.

[79] P. A. Wolf, R. D. Abbott, and W. B. Kannel, "Atrial fibrillation: a major contributor to stroke in the elderly. The Framingham Study," *Arch. Intern. Med.*, vol. 147, pp. 1561–1564, 1987. DOI: 10.1001/archinte.147.9.1561

[80] J. A. Cairns and S. J. Connolly, "Nonrheumatic atrial fibrillation. Risk of stroke and role of antithrombotic therapy," *Circulation*, vol. 84, pp. 469–481, 1991.

[81] M. Lamassa, A. Di Carlo, G. Pracucci, A. M. Basile, G. Trefoloni, P. Vanni, S. Spolveri, M. C. Baruffi, G. Landini, A. Ghetti, C. D. Wolfe, and D. Inzitari, "Characteristics, outcome, and care of stroke associated with atrial fibrillation in Europe: Data from a multicenter multinational hospital-based registry (The European Community Stroke Project)," *Stroke*, vol. 32, pp. 392–398, 2001.

[82] "Risk factors for stroke and efficacy of antithrombotic therapy in atrial fibrillation. Analysis of pooled data from five randomized controlled trials," *Arch. Intern. Med.*, vol. 154, pp. 1449–1457, 1994.

[83] R. G. Hart, O. Benavente, R. McBride, and L. A. Pearce, "Antithrombotic therapy to prevent stroke in patients with atrial fibrillation: A meta-analysis," *Ann. Intern. Med.*, vol. 131, pp. 492–501, 1999.

[84] R. G. Hart, L. A. Pearce, R. M. Rothbart, J. H. McAnulty, R. W. Asinger, and J. L. Halperin, "Stroke with intermittent atrial fibrillation: incidence and predictors during aspirin therapy. Stroke Prevention in Atrial Fibrillation Investigators," *J. Am. Coll. Cardiol.*, vol. 35, pp. 183–187, 2000. DOI: 10.1016/S0735-1097(99)00489-1

[85] The Stroke Prevention in Atrial Fibrillation Investigators, "Predictors of thromboembolism in atrial fibrillation: I. Clinical features of patients at risk," *Ann. Intern. Med.*, vol. 116, pp. 1–5, 1992.

[86] R. Farshi, D. Kistner, J. S. Sarma, J. A. Longmate, and B. N. Singh, "Ventricular rate control in chronic atrial fibrillation during daily activity and programmed exercise: A crossover open-label study of five drug regimens," *J. Am. Coll. Cardiol.*, vol. 33, pp. 304–310, 1999. DOI: 10.1016/S0735-1097(98)00561-0

[87] I. C. van Gelder, H. J. Crijns, R. G. Tieleman, J. Brugemann, P. J. de Kam, A. T. Gosselink, F. W. Verheugt, and K. I. Lie, "Chronic atrial fibrillation. Success of serial cardioversion therapy and safety of oral anticoagulation," *Arch. Intern. Med.*, vol. 156, pp. 2585–2592, 1996. DOI: 10.1001/archinte.156.22.2585

[88] D. Roy, M. Talajic, P. Dorian, S. Connolly, M. J. Eisenberg, M. Green, T. Kus, J. Lambert, M. Dubuc, P. Gagne, S. Nattel, and B. Thibault, "Amiodarone to prevent recurrence of atrial fibrillation. Canadian Trial of Atrial Fibrillation Investigators," *N. Engl. J. Med.*, vol. 342, pp. 913–920, 2000. DOI: 10.1056/NEJM200003303421302

[89] AFFIRM First Antiarrhythmic Drug Substudy Investigators, "Maintenance of sinus rhythm in patients with atrial fibrillation: An AFFIRM substudy of the first antiarrhythmic drug," *J. Am. Coll. Cardiol.*, vol. 42, pp. 20–29, 2003. DOI: 10.1016/S0735-1097(03)00559-X

[90] A. H. Madrid, M. G. Bueno, J. M. Rebollo, I. Marin, G. Pena, E. Bernal, A. Rodriguez, L. Cano, J. M. Cano, P. Cabeza, and C. Moro, "Use of irbesartan to maintain sinus rhythm in patients with long-lasting persistent atrial fibrillation: A prospective and randomized study," *Circulation*, vol. 106, pp. 331–336, 2002. DOI: 10.1161/01.CIR.0000022665.18619.83

[91] K. C. Ueng, T. P. Tsai, W. C. Yu, C. F. Tsai, M. C. Lin, K. C. Chan, C. Y. Chen, D. J. Wu, C. S. Lin, and S. A. Chen, "Use of enalapril to facilitate sinus rhythm maintenance after external cardioversion of long-standing persistent atrial fibrillation. Results of a prospective and controlled study," *Eur. Heart J.*, vol. 24, pp. 2090–2098, 2003. DOI: 10.1016/j.ehj.2003.08.014

[92] M. Ozaydin, E. Varol, S. M. Aslan, Z. Kucuktepe, A. Dogan, M. Ozturk, and A. Altinbas, "Effect of atorvastatin on the recurrence rates of atrial fibrillation after electrical cardioversion," *Am. J. Cardiol.*, vol. 97, pp. 1490–1493, 2006. DOI: 10.1016/j.amjcard.2005.11.082

[93] M. Haissaguerre, P. Jais, D. C. Shah, A. Takahashi, M. Hocini, G. Quiniou, S. Garrigue, A. Le Mouroux, P. Le Metayer, and J. Clementy, "Spontaneous initiation of atrial fibrillation by ectopic beats originating in the pulmonary veins," *New Eng. J. Med.*, vol. 339, pp. 659–666, 1998. DOI: 10.1056/NEJM199809033391003

[94] A. Natale, A. Raviele, T. Arentz, H. Calkins, S. A. Chen, M. Haissaguerre, G. Hindricks, Y. Ho, K. H. Kuck, F. Marchlinski, C. Napolitano, D. Packer, C. Pappone, E. N. Prystowsky, R. Schilling, D. Shah, S. Themistoclakis, and A. Verma, "Venice Chart international consensus document on atrial fibrillation ablation," *J. Cardiovasc. Electrophysiol.*, vol. 18, pp. 560–580, 2007. DOI: 10.1111/j.1540-8167.2007.00816.x

[95] R. Cappato, H. Calkins, S. A. Chen, W. Davies, Y. Iesaka, J. Kalman, Y. H. Kim, G. Klein, D. Packer, and A. Skanes, "Worldwide survey on the methods, efficacy, and safety of catheter ablation for human atrial fibrillation," *Circulation*, vol. 111, pp. 1100–1105, 2005. DOI: 10.1161/01.CIR.0000157153.30978.67

[96] K. Nademanee, J. McKenzie, E. Kosar, M. Schwab, B. Sunsaneewitayakul, T. Vasavakul, C. Khunnawat, and T. Ngarmukos, "A new approach for catheter ablation of atrial fibrillation: Mapping of the electrophysiologic substrate," *J. Amer. Coll. Card.*, vol. 43, pp. 2044–2053, 2004. DOI: 10.1016/j.jacc.2003.12.054

[97] K. T. Konings, J. L. Smeets, O. C. Penn, H. J. Wellens, and M. A. Allessie, "Configuration of unipolar atrial electrograms during electrically induced atrial fibrillation in humans," *Circulation*, vol. 95, pp. 1231–1241, 1997.

[98] P. Sanders, O. Berenfeld, M. Hocini, P. Jais, R. Vaidyanathan, L. F. Hsu, S. Garrigue, Y. Takahashi, M. Rotter, F. Sacher, C. Scavee, R. Ploutz-Snyder, J. Jalife, and M. Haissaguerre, "Spectral analysis identifies sites of high-frequency activity maintaining atrial fibrillation in humans," *Circulation*, vol. 112, pp. 789–797, 2005. DOI: 10.1161/CIRCULATIONAHA.104.517011

[99] D. Husser, M. Stridh, L. Sörnmo, P. Platanov, S. B. Olsson, and A. Bollmann, "Analysis of the surface electrocardiogram for monitoring and predicting antiarrhythmic drug effects in atrial fibrillation," *Cardiovasc. Drugs Therapy*, vol. 18, pp. 377–386, 2004. DOI: 10.1007/s10557-005-5062-z

[100] D. Husser, M. Stridh, L. Sörnmo, C. Geller, H. U. Klein, S. B. Olsson, and A. Bollmann, "Time–frequency analysis of the surface electrocardiogram for monitoring antiarrhythmic drug effects in atrial fibrillation," *Am. J. Cardiol.*, vol. 95, pp. 526–528, 2005. DOI: 10.1016/j.amjcard.2004.10.025

[101] D. Husser, M. Stridh, D. S. Cannom, A. K. Bhandari, M. J. Girsky, S. Kang, L. Sörnmo, S. B. Olsson, and A. Bollmann, "Validation and clinical application of time–frequency analysis of atrial fibrillation electrocardiograms," *J. Cardiovasc. Electrophysiol.*, vol. 18, pp. 41–46, 2007. DOI: 10.1111/j.1540-8167.2006.00683.x

[102] A. Bollmann, N. Kanuru, K. McTeague, P. Walter, D. B. DeLurgio, and J. Langberg, "Frequency analysis of human atrial fibrillation using the surface electrocardiogram and its response to ibutilide," *Am. J. Cardiol.*, vol. 81, pp. 1439–1445, 1998. DOI: 10.1016/S0002-9149(98)00210-0

[103] A. Bollmann, K. Binias, I. Toepffer, J. Molling, C. Geller, and H. Klein, "Importance of left atrial diameter and atrial fibrillatory frequency for conversion of persistent atrial fibrillation with oral flecainide," *Am. J. Cardiol.*, vol. 90, pp. 1011–1014, 2002. DOI: 10.1016/S0002-9149(02)02690-5

[104] A. Fujiki, T. Tsuneda, M. Sugao, K. Mizumaki, and H. Inoue, "Usefulness and safety of bepridil in converting persistent atrial fibrillation to sinus rhythm," *Am. J. Cardiol.*, vol. 92, pp. 472–475, 2003. DOI: 10.1016/S0002-9149(03)00672-6

[105] R. A. Schwartz and J. J. Langberg, "Atrial electrophysiological effects of ibutilide infusion in humans," *Pacing Clin. Electrophysiol.*, vol. 23, pp. 832–836, 2000. DOI: 10.1111/j.1540-8159.2000.tb00851.x

[106] A. Bollmann, M. Mende, A. Neugebauer, and D. Pfeiffer, "Atrial fibrillatory frequency predicts atrial defibrillation threshold and early arrhythmia recurrence in patients undergoing internal cardioversion of persistent atrial fibrillation," *Pacing Clin. Electrophysiol.*, vol. 25, pp. 1179–1184, 2002. DOI: 10.1046/j.1460-9592.2002.01179.x

[107] J. J. Langberg, J. C. Burnette, and K. K. McTeague, "Spectral analysis of the electrocardiogram predicts recurrence of atrial fibrillation after cardioversion," *J. Electrocardiol.*, vol. 31, pp. 80–84, 1998. DOI: 10.1016/S0022-0736(98)90297-7

[108] A. Bollmann, D. Husser, R. Steinert, M. Stridh, L. Sörnmo, S. Olsson, D. Polywka, J. Molling, C. Geller, and H. Klein, "Echo- and electrocardiographic predictors for atrial fibrillation recurrence following cardioversion," *J. Cardiovasc. Electrophysiol.*, vol. 14, pp. 162–165, 2003. DOI: 10.1046/j.1540.8167.90306.x

[109] F. Holmqvist, M. Stridh, J. E. Waktare, L. Sörnmo, S. B. Olsson, and C. J. Meurling, "Atrial fibrillatory rate and sinus rhythm maintenance in patients undergoing cardioversion of persistent atrial fibrillation," *Eur. Heart J.*, vol. 27, pp. 2201–2207, 2006. DOI: 10.1093/eurheartj/ehl098

[110] I. Kubara, H. Ikeda, T. Hiraki, T. Yoshida, M. Ohga, and T. Imaizumi, "Dispersion of filtered P wave duration by P wave signal-averaged ECG mapping system: Its usefulness for determining efficacy of disopyramide on paroxysmal atrial fibrillation," *J. Cardiovasc. Electrophysiol.*, vol. 10, pp. 670–679, 1999. DOI: 10.1111/j.1540-8167.1999.tb00244.x

[111] T. Yamada, M. Fukunami, T. Shimonagata, K. Kumagai, S. Sanada, H. Ogita, Y. Asano, M. Hori, and N. Hoki, "Dispersion of signal-averaged P wave duration on precordial body surface in patients with paroxysmal atrial fibrillation," *Eur. Heart J.*, vol. 20, pp. 211–220, 1999. DOI: 10.1053/euhj.1998.1281

[112] P. J. Stafford, J. Cooper, D. P. de Bono, R. Vincent, and C. J. Garratt, "Effect of low dose sotalol on the signal averaged P wave in patients with paroxysmal atrial fibrillation," *Br. Heart J.*, vol. 74, pp. 636–640, 1995. DOI: 10.1136/hrt.74.6.636

[113] W. Banasiak, A. Telichowski, S. D. Anker, A. Fuglewicz, D. Kalka, W. Molenda, K. Reczuch, J. Adamus, A. J. Coats, and P. Ponikowski, "Effects of amiodarone on the P-wave triggered signal-averaged electrocardiogram in patients with paroxysmal atrial fibrillation and coronary artery disease," *Am. J. Cardiol.*, vol. 83, pp. 112–114, 1999. DOI: 10.1016/S0002-9149(98)00792-9

[114] G. Opolski, P. Scislo, J. Stanislawska, A. Gorecki, R. Steckiewicz, and A. Torbicki, "Detection of patients at risk for recurrence of atrial fibrillation after successful electrical cardioversion by signal-averaged P-wave ECG," *Int. J. Cardiol.*, vol. 60, pp. 181–185, 1997. DOI: 10.1016/S0167-5273(97)02982-3

[115] K. Aytemir, S. Aksoyek, A. Yildirir, N. Ozer, and A. Oto, "Prediction of atrial fibrillation recurrence after cardioversion by P wave signal-averaged electrocardiography," *Int. J. Cardiol.*, vol. 70, pp. 15–21, 1999. DOI: 10.1016/S0167-5273(99)00038-8

[116] M. H. Raitt, K. D. Ingram, and S. M. Thurman, "Signal-averaged P wave duration predicts early recurrence of atrial fibrillation after cardioversion," *Pacing Clin. Electrophysiol.*, vol. 23, pp. 259–265, 2000. DOI: 10.1111/j.1540-8159.2000.tb00808.x

[117] M. Budeus, M. Hennersdorf, C. Perings, and B. E. Strauer, "The prediction of atrial fibrillation recurrence after electrical cardioversion with P wave signal averaged EKG," *Z. Kardiol.*, vol. 93, pp. 474–478, 2004.

[118] U. Dixen, C. Joens, J. Parner, V. Rasmussen, S. M. Pehrson, and G. B. Jensen, "Prolonged signal-averaged P wave duration after elective cardioversion increases the risk of recurrent atrial fibrillation," *Scand. Cardiovasc. J.*, vol. 38, pp. 147–151, 2004. DOI: 10.1080/14017430410028645

[119] F. Lombardi, A. Colombo, B. Basilico, R. Ravaglia, M. Garbin, D. Vergani, P. M. Battezzati, and C. Fiorentini, "Heart rate variability and early recurrence of atrial fibrillation after electrical cardioversion," *J. Am. Coll. Cardiol.*, vol. 37, pp. 157–162, 2001. DOI: 10.1016/S0735-1097(00)01039-1

[120] E. M. Kanoupakis, E. G. Manios, H. E. Mavrakis, M. D. Kaleboubas, F. I. Parthenakis, and P. E. Vardas, "Relation of autonomic modulation to recurrence of atrial fibrillation following cardioversion," *Am. J. Cardiol.*, vol. 86, pp. 954–958, 2000. DOI: 10.1016/S0002-9149(00)01129-2

[121] M. P. van den Berg, T. van Noord, J. Brouwer, J. Haaksma, D. J. van Veldhuisen, H. J. Crijns, and I. C. van Gelder, "Clustering of RR intervals predicts effective electrical cardioversion for atrial fibrillation," *J. Cardiovasc. Electrophysiol.*, vol. 15, pp. 1027–1033, 2004. DOI: 10.1046/j.1540-8167.2004.03686.x

[122] S. B. Olsson, N. Cai, M. Dohnal, and K. K. Talwar, "Noninvasive support for and characterization of multiple intranodal pathways in patients with mitral valve disease and atrial fibrillation," *Eur. Heart J.*, vol. 7, pp. 320–333, 1986.

[123] S. Rokas, S. Gaitanidou, S. Chatzidou, N. Agrios, and S. Stamatelopoulos, "A noninvasive method for the detection of dual atrioventricular node physiology in chronic atrial fibrillation," *Am. J. Cardiol.*, vol. 84, pp. 1442–1445, 1999. DOI: 10.1016/S0002-9149(99)00593-7

[124] S. Rokas, S. Gaitanidou, S. Chatzidou, C. Pamboucas, D. Achtipis, and S. Stamatelopoulos, "Atrioventricular node modification in patients with chronic atrial fibrillation: Role of morphology of RR interval variation," *Circulation*, vol. 103, pp. 2942–2948, 2001.

[125] A. M. Climent, M. S. Guillem, D. Husser, F. J. Castells, J. Millet, and A. Bollmann, "3-D Poincaré plot profiles—A novel non-invasive method to detect preferential ventricular response during atrial fibrillation," in *Proc. Comput. Cardiol.*, vol. 34, pp. 585–588, http://cinc.mit.edu, 2007.

[126] J. Tebbenjohanns, B. Schumacher, T. Korte, M. Niehaus, and D. Pfeiffer, "Bimodal RR interval distribution in chronic atrial fibrillation: Impact of dual atrioventricular nodal physiology on long-term rate control after catheter ablation of the posterior atrionodal input," *J. Cardiovasc. Electrophysiol.*, vol. 11, pp. 497–503, 2000. DOI: 10.1111/j.1540-8167.2000.tb00001.x

[127] A. Yamada, J. Hajano, S. Sakata, A. Okada, S. Mukai, N. Ohte, and G. Kimura, "Reduced ventricular response irregularity is associated with increased mortality in patients with chronic atrial fibrillation," *Circulation*, vol. 102, pp. 300–306, 2000.

CHAPTER 2

Time Domain Analysis of Atrial Fibrillation

Simona Petrutiu, Alan Sahakian, and Steven Swiryn

2.1 INTRODUCTION

The surface ECG characteristics are a direct reflection of pathophysiologic events in the atrium and can be used in studying atrial fibrillation (AF). By analyzing the ECG through a variety of signal processing techniques researchers have found clinically useful information that can be used to better understand and treat AF. Some of the advantages of using the ECG include the ability to record data for a long period of time, the minimal risks involved compared with invasive electrophysiologic study, and the ECG's reflection of the global activity in the atria and ventricles during AF.

The ECG can be used for manual or automatic detection of AF [1]. It directly reflects the electrophysiological processes that underlie AF, including refractory periods [2], autonomic tone [3], drug effects [4], and linking [5]. Therefore, it can be used to better understand the mechanisms and to study the effects of remodeling and the response to treatment with drugs or ablation during AF in humans. The ECG may also be used to predict the pattern of occurrence of AF, the likelihood of termination or persistence, and the probability of recurrence in different patients.

This chapter describes time domain methods used to characterize the ECG during AF. The electrocardiographic characteristics of AF will be presented as well as some methods commonly used to obtain and characterize the atrial activity. These methods include separating atrial from ventricular activity and obtaining a residual ECG in which only atrial activity is present, investigating the spatial and temporal organization of AF, and analyzing the residual ECG to learn about potential clinical uses. Fibrillatory wave characteristics will be discussed, including amplitude and rate, as well as the reproducibility of these characteristics over time. Frequency analysis is often used in addition to time domain methods to determine atrial rate and to characterize different patterns of occurrence of AF and the effects of drugs. Vector analysis, which can be used to study the organization of the arrhythmia, will be described as well as several alternative lead sets optimized for recording the atrial activity during AF.

2.2 ATRIAL FIBRILLATION AND ECG CHARACTERISTICS

In the ECG, AF is characterized by rapid atrial activity that is irregular in timing and morphology. Discrete P waves are absent and replaced by an oscillating baseline that consists of low-amplitude fibrillatory f-waves. The shape, amplitude, and regularity of f-waves vary from patient to patient. In some cases the pattern of atrial activity can be similar to atrial flutter, mostly regular and with high amplitude f-waves, while in other cases it can be less regular, have lower amplitude, or both. Atrial rates detected from the ECG in AF vary between 240 and 540 "beats" per minute (bpm) [6] with an average of 350 bpm, with even slower rates in the presence of anti-fibrillatory drugs and faster rates in the presence of acetylcholine. There is substantial overlap between rates during atrial flutter and AF. Distinguishing between AF and atrial flutter is clinically important since the treatment may be different for the two.

During AF, the RR intervals, and hence the ventricular rate, is commonly irregular. However, ventricular activity of this sort is present not only in AF, but also in a variety of other arrhythmias, including multifocal atrial tachycardia, atrial flutter with variable atrioventricular (AV) block, frequent premature atrial complexes, and sinus arrhythmia. Conversely, AF may be present with a regular ventricular rate, as in the case of AV block with artificial ventricular pacing. Although irregular ventricular activity is commonly associated with AF, it often fails as the sole diagnostic criterion. In fact, the presence of AF has been shown to be under-recognized in paced patients, with important adverse clinical consequences since the RR intervals are regular [7]. Figure 2.1 illustrates some examples of AF and atrial flutter with irregular and regular ventricular rhythms.

The characteristics of the RR intervals vary from the onset of AF to its termination. The variation present during AF is much larger than the variations typically present in sinus rhythm. Figure 2.2 shows an example of RR intervals over 25 min of sinus rhythm and 25 min of AF, and the larger variability during AF can be observed. Since paroxysmal AF involves episodes of AF which self-terminate, it is reasonable to analyze the RR intervals of these entire episodes. Gallagher et al. observed that the mean RR interval and the variability of the RR intervals increase from the onset [8]. In addition, they found that immediately before termination, the duration and variability of the RR intervals increase. This increase may indicate increased parasympathetic influence at the AV node. Another possible interpretation of the change in the pattern of the ventricular response before termination may be related to changes occurring in the atria prior to termination.

2.3 FIBRILLATORY WAVE ANALYSIS

The irregularity of the ventricular response can be used in the detection of AF, but as previously described it is neither 100% specific nor 100% sensitive for AF. In order to better understand the mechanisms of AF and the effects of drugs or autonomic tone on the arrhythmia, it is necessary to study the characteristics of the atrial fibrillatory waves themselves.

The analysis of atrial activity in the ECG is complicated by the simultaneous presence of ventricular activity which is typical of much larger amplitude. There are two approaches to investigate

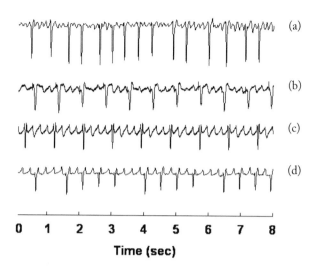

Figure 2.1: Atrial fibrillation and atrial flutter examples. (a) AF with irregular ventricular rhythm, (b) AF with paced (regular) ventricular rhythm, (c) atrial flutter with regular ventricular rhythm, (d) atrial flutter with irregular ventricular rhythm.

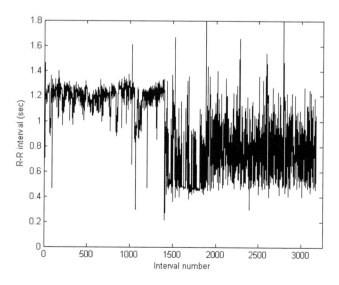

Figure 2.2: RR interval variability over 50 min the first 1,400 intervals correspond to sinus rhythm and the remaining correspond to AF. The variability is much larger and the RR intervals are shorter for AF compared to sinus rhythm. These interval properties help in the detection of AF.

f-waves: isolate segments free of QRST complexes [9] where only f-waves are present or cancel the ventricular activity and obtain a residual ECG that consists only of atrial activity [10, 11, 12, 13, 14]; the latter approach is described in detail in Chapter 3.

A study by Xi et al. validated different cancellation methods by comparing gold standard 'pure' AF ECGs obtained in patients with complete AV block after undergoing AV junctional ablation by briefly stopping pacing with residual ECGs obtained by cancellation of paced beats before and after the pure AF segment [15]. Comparison between pure AF segments and QRST cancelled segments preceding and following the pure AF segments showed similar characteristics in the frequency domain.

Leads II and V1 have the largest ratio of atrial to ventricular signal amplitude and are therefore often chosen for analysis. It is believed that subtraction has the best results in lead V1, resulting in high amplitude f-waves, and this is due to the lead's proximity to the right atrium.

After obtaining a residual ECG, f-wave characterization can be accomplished. The parameters to be analyzed include f-wave amplitude and rate. The distribution of amplitude over time can also be quantified using the amplitude probability density function, as is done for the intra-atrial leads [6]. Frequency domain analysis is often used in addition to time domain analysis to determine atrial rate and to characterize different patterns of occurrence of AF and the effects of drugs.

2.4   FIBRILLATORY WAVE AMPLITUDE

In the ECG, f-wave amplitude may be observed by direct measurement or by examination of the residual ECG. The f-waves may be classified as "fine" or "coarse," i.e., as less than or greater than 50 μV [16]. The amplitude may be further described as "very coarse," i.e., larger than 0.25 mV, or "straight line," i.e., indistinguishable from the baseline [17].

The f-wave amplitude may be measured in a number of ways. One such method involves identifying the four largest peak-to-peak amplitude f-waves in the residual ECG [18]. The four f-waves are selected outside the QRST regions to avoid QRST residuals. The average amplitude of the four f-waves is considered to be representative. Another method for finding the f-wave amplitude involves averaging the f-wave amplitudes for the entire residual ECG. Upon comparison, these two methods yield almost identical results.

Characterization of f-wave amplitude implies stability over time. Xi et al. found that f-wave amplitude measurements were repeatable for multiple ECGs taken from the same patient over 24 h; interpatient variability, however, was substantial [19]. They also determined that age, use of anti-fibrillatory drugs, and AF pattern (paroxysmal, persistent, or permanent) did not cause statistically significant differences in f-wave amplitude [18]. Takahashi et al. also investigated the relationship between f-wave amplitude and AF pattern, but with different results [20]. They found that f-waves were smaller during paroxysmal AF than in persistent AF, but that the presence of larger f-waves in paroxysmal AF indicated a higher likelihood of progression to persistent AF.

Researchers have made various attempts to relate f-wave amplitude to heart disease; their efforts have produced conflicting results. As early as 1915, Hewlett and Wilson observed an AF patient with larger f-waves, but did not draw any conclusions about this finding [21]. In 1930, Cookson reported that f-waves were 'medium to large' in patients with rheumatic heart disease and 'medium to small' in patients with other kinds of heart disease [22]. In 1962, Thurmann and Janney examined the relationship between AF coarseness and rheumatic versus arteriosclerotic heart disease [16]. They found the majority of patients with coarse AF in lead V1 had rheumatic heart disease and the majority of patients with fine AF in lead V1 had arteriosclerotic heart disease. In 1963, the work of Culler et al. supported these results by showing that coarse f-waves were prevalent in rheumatic heart disease and fine f-waves were prevalent in arteriosclerotic heart disease [17]. Culler et al. also found that coarse f-waves were related to left atrial enlargement when rheumatic heart disease was present. In 1979, Morganroth et al. refuted previous findings of relationships between f-wave amplitude and left atrial size and between f-wave amplitude and the etiology of heart disease [23]. Previous studies had used roentgenography, surgery, or autopsy to find left atrial size, which may have resulted in a bias toward rheumatic heart disease; at the time, surgeries were only performed for rheumatic heart disease. Morganroth et al., on the other hand, used echocardiography to measure left atrial size which may have produced more clinically relevant findings. In 1988, Aysha and Hassan also used echocardiography to study the relationship between average f-wave amplitude and left atrial size [24]. They found, however, that average f-wave amplitude and left atrial size were significantly larger in patients with chronic and rheumatic AF than in patients with paroxysmal and chronic nonrheumatic AF. They also found that f-wave amplitude was highly correlated with left atrial size. Due to the conflicting results of these and other studies, no overall consensus exists as to the relationship between f-wave amplitude and the etiology of heart disease.

Additional relationships involving f-wave amplitude have also been investigated. Bollmann et al. studied the relationship between AF coarseness and left atrial appendage flow velocity; they concluded that the relationship depends upon the recording technique used, i.e., orthogonal versus standard ECG leads [25]. Bollmann et al. maintained that fibrillatory activity during AF can be attributed in part to the mechanical contractions of the left atrial appendage. Blackshear et al. concluded that f-wave amplitude does not correlate with left atrial appendage velocity, left atrial size, increased left ventricular mass, systolic dysfunction, hypertension, or risk of systemic embolism [26].

2.5   FIBRILLATORY WAVE RATE

Although AF is, by definition, not a completely regular atrial rhythm such as atrial flutter or certain atrial tachycardias, a quantifiable level of temporal organization exists [27]. Slocum et al. used the autocorrelation function to quantify signal organization and estimate atrial rate. This method was successful in discriminating between AF and AV-dissociated P waves from the residual ECG [28]. The autocorrelation of the residual signal contains peaks at the atrial cycle length and multiples of the cycle length, if the atrial activity is periodic and AV-dissociated.

The discrimination between AF and AV-dissociated P waves was based on the peak amplitude and duration of the autocorrelation function in leads II and V1. The best parameter for detecting AF seemed to be peak amplitude with a significantly higher peak for AF recordings compared to the control group. The best detection algorithm was achieved when peak amplitudes in leads II and V1 were added, and this new parameter was used as a classifier. The properties of AF which allow estimation of atrial rate using autocorrelation can also be exploited in the frequency domain and are reflected by the narrowband characteristics of the power spectrum of the residual ECG.

It has become increasingly apparent that the best way to investigate the ECG during AF is by using a combination of time and frequency domain methods. The steps of generating the residual ECG are performed in the time domain, but additional information can be obtained by further investigation of the residual f-waves in the frequency domain; this will be further described in Chapter 4.

2.6   REPRODUCIBILITY OVER TIME

Fibrillatory waves on the ECG have been scrutinized to search for the reflection of different underlying mechanisms, the effects of electrophysiologic and structural remodeling, and the response to different drug therapies. There has been an underlying premise that f-wave characteristics in individual patients do not vary randomly and are constant during stable clinical conditions.

This premise has been tested in a study by Xi et al. where a series of 10 standard ECGs were recorded during 24 h in a group of 20 clinically stable patients with AF [19]. After QRST cancellation, the f-waves from the residual ECG where analyzed to investigate interpatient versus intrapatient differences. Parameters such as peak-to-peak amplitude and short-term peak frequency were evaluated by analysis of variance (ANOVA). Peak-to-peak amplitude ranged from 0.06–0.35 mV, and one standard deviation of the amplitude for each patient ranged from 0.004–0.053 mV. Short-term peak frequencies ranged from 4.6–8.0 Hz, and one standard deviation for each patient ranged from 0.2–0.5 Hz. Interpatient differences were significantly higher compared to intrapatient differences for these parameters. This demonstrated that f-wave characteristics are repeatable from ECG to ECG over 24 h for clinically stable patients, but varied greatly from patient to patient.

This study implies that a standard 10-s ECG provides a long enough duration to provide initial characterization of AF. However, the characteristics of f-waves were not addressed in a longer time frame (longer than 24 h) or with respect to a particular type of AF.

A subsequent study investigated whether f-wave characteristics reflect the pattern of AF occurrence [18]. Clinically, patients are sometimes characterized according to three patterns of occurrence of AF: paroxysmal, persistent, and permanent. This classification is important because different management strategies are required based on the type of AF.

As previously described, the peak in the power spectrum is a direct reflection of atrial rate. Fibrillatory waves were found to reflect specific clinical variables, with higher frequency in permanent than in paroxysmal fibrillation, but lower frequency in older than in younger patients. Beta blockers were associated with lower amplitude and decreased f-wave rate.

2.7   SUITABLE LEAD SETS FOR AF ANALYSIS

Standard 12-lead ECG systems may not be optimal for studying atrial activity. More specifically, the information extracted from the ECG about the atrial activity during AF may be limited because of the small number of electrodes used and their location. The availability of a much larger number of electrodes, as in the case of body surface potential mapping (BSPM), increases the information content, however, the availability of BSPM systems in a clinical setting is limited and not very practical. Therefore, several studies have investigated better possibilities of examining the atrial activity during AF, while using standard equipment and placing a small number of electrodes in optimal positions.

One of the proposed systems is named the atriocardiogram (ACG) [29]. This adapted system maintains the same number of electrodes as the standard 12-lead ECG in order to reduce the complexity of lead placement during the current clinical routine. Of the nine electrodes involved in recording the standard 12-lead ECG, the limb lead electrodes are left in place, as well as the precordial electrodes V1 and V2. Electrodes sensing V3 to V6 are repositioned in a counterclockwise fashion around those of V1 and V2. The precordial electrodes are arranged to form a 2×3 grid of the upper right chest lying over the atria. V3 is placed one intercostal space above V2 and V4 one intercostal space above V1. V5 is repositioned at the right of the new V4 position and V6 below the new V5 position. The placement is illustrated in Figure 2.3.

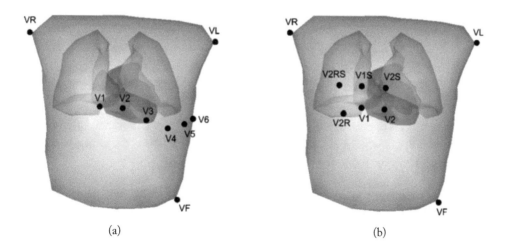

(a)   (b)

Figure 2.3: The atriocardiogram lead system. Display of biophysical model showing geometries of the thorax, lungs, atria, ventricles, and blood filled cardiac cavities [30]. (a) The standard 12-lead ECG, and (b) the atriocardiogram lead system. The electrode positions are indicated by black dots.

This work has subsequently been extended to find the optimal modified 12-lead system which is referred to as the optimal ACG (OACG) [30]. Five of the nine electrodes of the standard 12-lead

ECG are again left in place, the limb leads and two precordial leads electrodes, V1 and V4. Lead V1 was selected since it has the most proximal location to the right atrium. Lead V4 was selected because it was found to have the lowest correlation with lead V1, therefore providing the maximally independent view of AF. The other four electrode positions were found by searching 64 nodes on the thorax. Electrode V1S (V1 superior) is placed one intercostal space above the V1 electrode position. Electrode V2RS is placed at the right of V1S at the same height. The third electrode VLC is positioned just below the left clavicle. The last electrode V1P (V1 posterior) was positioned on the back just behind the atria at the same level as V1. It can be noted that the first two positions were also included in the ACG lead system. This OACG lead system is presented in Figure 2.4.

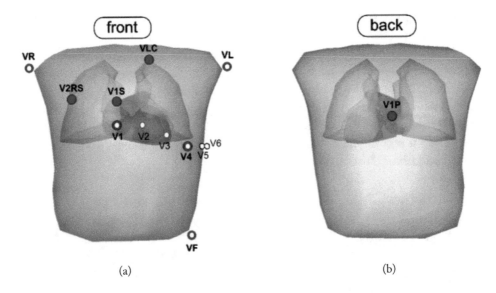

(a)   (b)

Figure 2.4: The optimal atriocardiogram lead system viewed from (a) the front and (b) the back. The white dots indicate the standard 12-lead ECG electrode positions, whereas the larger, heavy dots are electrode locations of the proposed lead system.

It has been shown that the left atrium and the pulmonary veins (PVs) play an important role in the initiation and maintenance of AF [31]. In many patients, rapid focal firing within the PVs may trigger the onset of AF. Recent studies have demonstrated a gradient of frequencies between the left and right atria of patients in AF undergoing PV isolation [32]. Higher frequencies were observed in the left atrium when compared to the right atrium in most patients. The presence of a frequency gradient is indicative of the type of AF and improves the likelihood of success of the ablation procedure.

The standard ECG leads present more detail about ventricular activity, but are especially limited in recording left atrial activity. Several tools have been developed for quantitative analysis of atrial f-waves from the ECG as well as intracardiac recordings [11, 12, 13, 33, 34]. During AF, most techniques of investigating atrial f-waves from the ECG involve analyzing lead V1 because it has the largest ratio of atrial amplitude compared to ventricular amplitude [34]. Since this lead is in close proximity to the right atrium it is thought of as reflecting mostly right atrial activity. By comparing the information extracted from the ECG with simultaneous intracardiac recordings, Petrutiu et al. showed that local events can be characterized from the ECG and that the infrequently used posterior leads seem to better reflect left atrial events than the standard 12-leads currently used in clinical practice [35].

In addition to the standard ECG limb leads and the precordial leads V1, V2, the standard but infrequently used leads V7, V8, and V9 were recorded. The V7–V9 electrodes extend in a horizontal line from V6. The V7 electrode is placed at the posterior axillary line, V8 at the level of V7 at the mid-scapular line, and V9 at the level of V8 at the paravertebral line. An additional surface lead location was defined, V10, which was also placed on the paravertebral line, above V9, on the same level as V1.

Intra-atrial recordings from the right and left atrium were obtained simultaneously with the ECG. The ECG signals were processed to isolate f-waves by using a template matching algorithm similar to the one originally described by Slocum et al. and to obtain a residual ECG [1]. Following QRST cancellation, the power spectrum of each residual ECG was calculated using the Fast Fourier Transform (FFT). The peak frequency was determined in each lead as the location where the maximum peak of the power spectrum occurred. Dominant frequencies (DFs) were obtained from intra-atrial electrograms and compared to the peak frequencies obtained from the different ECG leads. A strong relationship was observed between the ECG peak frequency and the intra-atrial recording DF when the two were close in location. Correlation was found to be the highest between V1 and the right atrium recording, while V9 had a close correlation with the left atrium recording. When recording from locations that were far away from each other, the agreement between the two measurements decreased. This demonstrate that local atrial events and interatrial frequency gradients can be characterized from the ECG. Lead V1 reflects mostly right atrial activity, while V9 reflects mostly left atrial activity.

It has been shown that different alternatives of electrode positioning exist that have better performance when recording the atrial activity during AF. These adapted systems use the same number of electrodes as the standard 12-lead ECG in order to be easily integrated during the current clinical routine. The infrequently used posterior leads and other more optimal lead locations were found to improve the information content available about the atrial activity during AF and were superior to the standard 12-leads currently used.

2.8   VECTOR ANALYSIS

During sinus rhythm, the morphology of P waves in the 12-lead ECG can offer important physiologic and pathophysiologic information. For example, the P waves in V1, in the normal heart, are usually biphasic in morphology with a positive deflection followed by a negative deflection. A positive deflection with significantly larger amplitude than the negative deflection can indicate right atrial enlargement, while larger negative deflection can indicate left atrial enlargement [36].

Sometimes an atrial beat can originate from a location other than the sinus node. These 'ectopic' beats usually occur earlier than the next sinus beat would occur and have the potential to induce an atrial arrhythmia. In addition to an earlier timing, an atrial beat of ectopic origin can be identified by a different P wave morphology than that of sinus rhythm. A P wave of a premature atrial beat will look more similar to a P wave of sinus origin if the ectopy has a right atrial origin than if it were a left atrial origin [37].

Characteristics in the ECG can also often identify the mechanisms of atrial flutter, which is an arrhythmia usually consisting of a reentrant circuit around an anatomic obstacle that rapidly drives the rest of the atria. The most common form of atrial flutter is a circuit around the tricuspid annulus. This particular type of atrial flutter is often successfully treated by radiofrequency ablation of a narrow isthmus between the tricuspid annulus and the inferior vena cava, which is a critical pathway for the circuit [38]. The atrial activity on the ECG is characterized by a continuous waveform that is regular in timing with a negative deflection in the inferior leads (II, III, aVF) and a positive deflection in V1. When the ECG shows continuous and regular atrial activity with morphologies unlike that of the typical flutter described above, they are often classified as 'atypical' atrial flutter and are likely to have a different circuit.

This brings us to the question of whether the morphology of the ECG during AF can convey similar information about the particular mechanisms of the arrhythmia, similar to how P waves during sinus rhythm, atrial ectopy, or atrial flutter can contain useful information. The challenge of analyzing atrial f-waves in this fashion is complicated by the lack of discrete waveforms that are characteristic of atrial beats of sinus node or ectopic origin, and the lack of regularity that is characteristic of atrial flutter. This complexity of the atrial fibrillatory signal is a reflection of the complexity of the activation wavefronts from the atria itself.

Ng et al. hypothesized that information about atrial activation during AF could be analyzed using vectorcardiography [39]. In vectorcardiography, three orthogonal leads are plotted against each other. The result is a series of arrows or "comets" that form a vector loop and track a sequence of activation directions for a cardiac cycle. The two- or three-dimensional activation direction at any given time can determined by the vector starting from the point of origin (no voltage) to a point on the loop, thus displaying what cardiologists typically reconstruct in their minds from the 12-lead ECG. The comets are spaced at constant time intervals, which are typically 10 ms, and provide a sense of how rapidly the vectors change. Therefore, any preferred activation directions or sequence of activation directions that may exist for AF might be evident from the analysis of vectorcardiograms of several atrial activation cycles.

Vectorcardiography has most commonly been used to analyze QRS complexes to identify types of ventricular conduction abnormalities such as bundle branch and fascicular block and axis deviation [40]. Because vectorcardiograms have not been rigorously tested for atrial activity or atrial arrhythmias, Ng et al. first sought to validate this technique on atrial flutter ECGs, where the mechanisms of the arrhythmia could be well-established before using this approach on AF [41]. As described above, the circuit for typical flutter is bounded by the tricuspid annulus. When facing the patient from an angle that is slightly on the patient's left (i.e., left-anterior oblique view), we would be looking directly at the opening of the tricuspid valve. From this view, the activation sequence of typical flutter is most commonly counterclockwise around the annulus. However, clockwise activation sequences are also possible and can be distinguished from counterclockwise sequences by intra-atrial mapping. After vector loops of the atrial flutter waves were created from the 12-lead ECG, the planes that best fit the vector loops were determined using principal component analysis.

An illustration of the process of obtaining the plane of best fit is shown in Figure 2.5. The principal axis is defined as the direction of the loop that contains the greatest variance. The direction

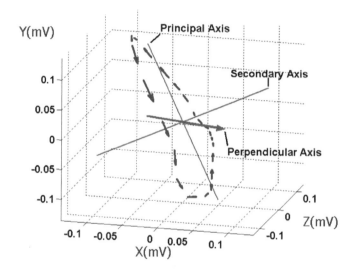

Figure 2.5: Obtaining the plane of best fit in the vectorcardiogram: the principal axis is defined as the direction of the loop that contains the largest variance and the secondary axis is chosen as the direction perpendicular to the primary axis that has the greatest variance.

perpendicular to the primary axis that has the largest variance is the secondary axis. The plane of best fit can therefore be defined by the vector perpendicular to the principal and secondary axes. The perpendicular axis of the plane of best fit was described by Ng et al., using the convention of azimuth and elevation angles [41]. Figure 2.6 demonstrates how the orientation of the planes relates to azimuth and elevation angles. The frontal plane was defined to have azimuth and elevation angles of zero degrees. Rotation of the plane to the right results in positive azimuth angles, while leftward rotation results in negative angles. An upward rotation results in positive elevation angles.

It was found that almost all the planes of best fit were approximately parallel to the planes where the tricuspid annulus would be expected to be [41]. Furthermore, the vector loops had rotations

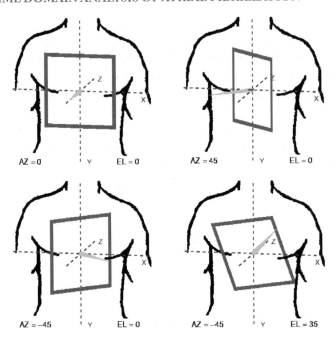

Figure 2.6: Orientation of the planes in relationship to azimuth and elevation angles. Frontal plane defined to have azimuth and elevation angles of zero degrees; rotation of the plane to the right results in positive azimuth angles; rotation of the plane to the left results in negative azimuth angles; upward rotation results in positive elevation angles.

corresponding to the respective counterclockwise and clockwise activation sequences of the atrial flutter as shown in Figure 2.7. Thus, the results from this study confirm that vector analysis could accurately reflect atrial activation directions and sequences during atrial tachyarrhythmias.

To study vector loops during AF, f-waves should preferably be obtained when ventricular waveforms are not present [39]. Because such intervals are generally short and cancellation techniques may introduce QRS residuals, ECGs from patients undergoing atrioventricular node ablation for pacemaker implantation were used for analysis. To test for block following the ablation, the pacemaker is turned off for two or three seconds during which the f-waves will be completely uncovered. The atrial cycle length of each of these segments was estimated using autocorrelation. Thus, the number of cycles analyzed was equal to the length of the segment divided by the estimated cycle length.

Vector loops were constructed from each f-wave cycle and the planes of best fit were determined as described for atrial flutter; see Figure 2.8. The azimuth angles of the perpendicular of the planes were categorized into twelve 30° bins. If AF activation has some organization with preferred activation directions, one would expect certain 30° bins with a large percentage of the planes. If not, they would be more evenly distributed. In their study, Ng et al. found that 15 of the 22 ECGs had

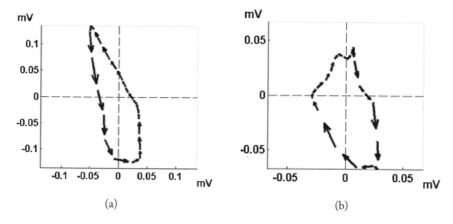

(a)   (b)

Figure 2.7: ECG atrial flutter waves viewed from the left anterior oblique perspective (a) with counter-clockwise and (b) clockwise reentrant circuits around the tricuspid annulus.

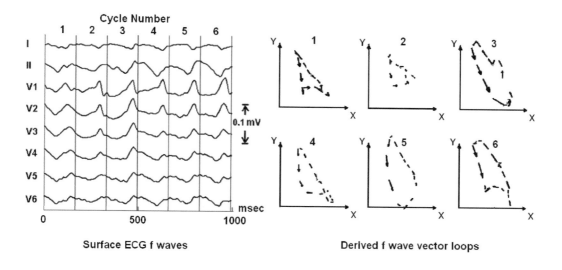

Figure 2.8: 1-s "pure"" AF signal divided into six segments of equal length and the three-dimensional vector loops of the individual estimated AF cycles.

at least 30% of the planes in a single 30° region of azimuth angles. Assuming that each bin had an equal chance of containing the plane of best fit, the probability of 30% of the planes in a single 30° range by chance was very low ($p = 0.03$, assuming the mean 18 fibrillatory wave cycles per episode). Of these 15 episodes, 12 had the highest percentage of planes with azimuth angles within 30° of the sagittal plane. This suggests that not only is organized activity commonly observed during AF, but these activation patterns are similar across many patients. Also, it appears that there are patients with episodes that are less organized, which suggest that there may be a range of complexity of AF organization.

The finding that atrial activations during AF had preferred activation sequences confirmed globally what Gerstenfeld et al. discovered locally with vector loops constructed from recordings from an orthogonal catheter placed in the right atrium [5]. Others have also demonstrated local AF organization through a variety of methods using epicardial and endocardial recordings [42, 43, 44].

Haissaguerre et al. showed that focal triggers of AF commonly come from the pulmonary veins [31]. From this finding, ablation therapies that isolate the PVs have been shown to be effective in preventing AF recurrences. Therefore, if the general activation pattern is going from the left to the right atrium, the consistent vector loops may be a reflection of that pattern. Additionally, atrial fiber orientations and the constant location of anatomic obstacles (most prominently the AV valves) must constrain atrial activation sequences during AF. Lazar et al. further showed that frequency gradients exist between the left and right atrium in patients with paroxysmal AF, further supporting the role of the left atrium as a driver [32]. In that study it was also shown that patients with persistent AF do not exhibit such gradients. Whether the difference in gradients between paroxysmal versus persistent is due to mechanisms, atrial substrate, or electrophysiologic remodeling is unknown. It is possible that the differences in vector organization may reflect the presence or absence of these gradients. However, this has not yet been studied.

Another possible explanation for the organized vector activity is that the vector loops reflect stable spiral waves or rotors that act as drivers for AF [45]. Jalife et al. were able to induce such rotors in animal models. In this model, mother rotors were stable high-frequency sources that broke up into daughter rotors away from the sources. If this model can be extrapolated to the human heart, then there may be potential to locate the mother rotor using vector analysis.

Although vector analysis has taught us much about AF dynamics, questions still remain regarding what the vector loops reflect. One question is whether vector loops derived from the 12-lead ECG reflect both right and left atrial activity. Analysis of atrial flutter suggests that right atrial events dominate in the 12-lead ECG. A true orthogonal lead set may be advantageous in this regard because the electrode placed in the back may better detect left atrial activity. High resolution BSPM may offer additional information of atrial activation during AF. Another question is how interatrial frequency gradients will be reflected in vector analysis. Frequency gradients are thought to be a result of high-frequency sources which drive the rest of the atria. It is unknown whether vectors will exhibit disorganized behavior due to the lack of coherence between regions with respect to frequency differences, or if the consistent activation direction from a stable source will produce organized vectors.

Our ability to understand the mechanisms of AF is limited by the difficulty to map the intact atria in their entirety with enough detail. Even recent studies that have recorded a thousand local electrograms could not cover all points in the right and left atrial endocardium, epicardium, and the interatrial septum. Thus, the ECG and vector analysis offer another piece of global information that can be used to supplement the information obtained from local electrograms to gain better understanding of this complicated arrhythmia.

2.9   LIMITATIONS OF AF ANALYSIS

One challenge in analyzing f-wave characteristics from the ECG is the low-amplitude of the atrial activity. The presence of artifact and noise also makes this more difficult. Since the ECG also reflects interpatient differences in body characteristics and is affected by other physiological parameters, it is difficult to understand the influence of pathophysiology on f-wave amplitude. This is reflected in reports of conflicting results when trying to relate f-wave amplitude with different etiologies of AF.

During AF, the atria are being activated in an uncoordinated and noncoherent manner. The signals recorded in the ECG are globally recorded signals and therefore it is difficult to infer much about local activations and spatial differences. Most time domain methods use lead V1 for analysis which is dominated by right atrial activity due to its proximal location to the right atrium. Therefore, it may be difficult to study the more important left atrial events during AF using the standard 12-lead ECG. However, new systems have been proposed which seem optimal for recording the atrial activity during AF. Intracardiac signals, typically obtained from catheters, exhibit the opposite problem of having a myopic field of view unless a large number of recording sites are used. Therefore, the ECG and intracardiac electrograms may be best used simultaneously in the study of AF mechanisms, as they provide complementary information.

2.10   CONCLUSIONS

The ECG during AF is not random, but provides a direct reflection of pathophysiologic mechanisms of AF and atrial anatomy. The mechanisms of AF, the effects of electrophysiological and structural remodeling, as well as the effectiveness of different treatments can be investigated from the ECG using time domain methods to characterize the atrial activity during AF. This chapter describes signal processing methods as well as observations that can be obtained directly from the ECG signal, such as the general characteristics of AF. These methods include cancellation techniques, autocorrelation, vector analysis, and the use of modified lead sets optimized for recording atrial activity. Whether the study of atrial activity in the ECG can be used to distinguish between different mechanisms of AF is not yet known, but further investigation can improve our understanding of these mechanisms and help with the management of this common arrhythmia.

Bibliography

[1] J. Slocum, A. V. Sahakian, and S. Swiryn, "Diagnosis of atrial fibrillation from surface electrocardiograms based on computer-detected atrial activity," *J. Electrocardiol.*, vol. 25, pp. 1–8, 1992. DOI: 10.1016/0022-0736(92)90123-H

[2] M. C. Wijffels, C. J. Kirchhof, R. Dorland, and M. A. Allessie, "Atrial fibrillation begets atrial fibrillation. A study in awake chronically instrumented goats," *Circulation*, vol. 92, pp. 1954–1968, 1995.

[3] M. Zimmermann, "Autonomic tone and atrial fibrillation," *J. Cardiovasc. Electrophysiol.*, vol. 14, pp. 559–564, 2003. DOI: 10.1046/j.1540-8167.2003.03156.x

[4] A. Bollmann, N. Kanuru, K. McTeague, P. Walter, D. B. DeLurgio, and J. Langberg, "Frequency analysis of human atrial fibrillation using the surface electrocardiogram and its response to ibutilide," *Am. J. Cardiol.*, vol. 81, pp. 1439–1445, 1998. DOI: 10.1016/S0002-9149(98)00210-0

[5] E. P. Gerstenfeld, A. V. Sahakian, and S. Swiryn, "Evidence for transient linking of atrial excitation during atrial fibrillation in humans," *Circulation*, vol. 86, pp. 375–82, 1992.

[6] J. Slocum, A. V. Sahakian, and S. Swiryn, "Computer discrimination of atrial fibrillation and regular atrial rhythms from intra-atrial electrograms," *Pacing Clin. Electrophysiol.*, vol. 11, pp. 610–621, 1988. DOI: 10.1111/j.1540-8159.1988.tb04557.x

[7] A. M. Patel, D. C. Westveer, K. C. Man, J. R. Stewart, and H. I. Frumin, "Treatment of underlying atrial fibrillation: paced rhythm obscures recognition," *J. Am. Coll. Cardiol.*, vol. 36, pp. 784–787, 2000. DOI: 10.1016/S0735-1097(00)00794-4

[8] M. M. Gallagher, K. Hnatkova, F. D. Murgatroyd, J. E. Waktare, X. Guo, A. J. Camm, and M. Malik, "Evolution of changes in the ventricular rhythm during paroxysmal atrial fibrillation," *Pacing Clin. Electrophysiol.*, vol. 21, pp. 2450–2454, 1998. DOI: 10.1111/j.1540-8159.1998.tb01199.x

[9] D. S. Rosenbaum and R. J. Cohen, "Frequency based measures of atrial fibrillation in man," in *Proc. IEEE EMBS*, vol. 12, pp. 582–583, 1990.

[10] J. Slocum, E. Byrom, L. McCarthy, A. V. Sahakian, and S. Swiryn, "Computer detection of atrioventricular dissociation from surface electrocardiograms during wide QRS complex tachycardia," *Circulation*, vol. 72, pp. 1028–1036, 1985.

[11] M. Stridh and L. Sörnmo, "Spatiotemporal QRST cancellation techniques for analysis of atrial fibrillation," *IEEE Trans. Biomed. Eng.*, vol. 48, pp. 105–111, 2001. DOI: 10.1109/10.900266

[12] P. Langley, J. P. Bourke, and A. Murray, "Frequency analysis of atrial fibrillation," in *Proc. Comput. Cardiol.*, vol. 27, pp. 65–68, IEEE Press, 2000. DOI: 10.1109/TBME.2004.827272

[13] J. J. Rieta, F. Castells, C. Sánchez, V. Zarzoso, and J. Millet, "Atrial activity extraction for atrial fibrillation analysis using blind source separation," *IEEE Trans. Biomed. Eng.*, vol. 51, pp. 1176–1186, 2004.

[14] P. Langley, M. Stridh, J. J. Rieta, J. Millet, L. Sörnmo, and A. Murray, "Comparison of atrial signal extraction algorithms in 12-lead ECGs with atrial fibrillation," *IEEE Trans. Biomed. Eng.*, vol. 53, pp. 343–346, 2006. DOI: 10.1109/TBME.2005.862567

[15] Q. Xi, A. V. Sahakian, and S. Swiryn, "The effect of QRS cancellation on atrial fibrillatory wave signal characteristics in the surface electrocardiogram," *J. Electrocardiol.*, vol. 36, pp. 243–249, 2003. DOI: 10.1016/S0022-0736(03)00046-3

[16] M. Thurmann and J. Janney, "The diagnostic importance of fibrillatory wave size," *Circulation*, vol. 25, pp. 991–994, 1962.

[17] M. R. Culler, J. A. Boone, and P. C. Gazes, "Fibrillatory wave size as a clue to etiological diagnosis," *Am. Heart J.*, vol. 66, pp. 435–436, 1963. DOI: 10.1016/0002-8703(63)90280-1

[18] Q. Xi, A. V. Sahakian, T. G. Frohlich, J. Ng, and S. Swiryn, "Relationship between pattern of occurrence of atrial fibrillation and surface electrocardiographic fibrillatory wave characteristics," *Heart Rhythm*, vol. 1, pp. 656–663, 2004. DOI: 10.1016/j.hrthm.2004.09.010

[19] Q. Xi, A. V. Sahakian, J. Ng, and S. Swiryn, "Atrial fibrillatory wave characteristics on surface electrocardiogram," *J. Cardiovasc. Electrophysiol.*, vol. 15, pp. 911–917, 2004.

[20] N. Takahashi, A. Seki, K. Imataka, and J. Fujii, "Fibrillatory wave size in paroxysmal atrial fibrillation," *Jpn. Heart J.*, vol. 24, pp. 309–314, 1983.

[21] B. Mutlu, M. Karabulut, E. Eroglu, K. Tigen, F. Bayrak, H. Fotbolcu, and Y. Basaran, "Fibrillatory wave amplitude as a marker of left atrial and left atrial appendage function, and a predictor of thromboembolic risk in patients with rheumatic mitral stenosis," *Int. J. Cardiol.*, vol. 91, pp. 179–186, 2003. DOI: 10.1016/S0167-5273(03)00024-X

[22] P. de Silva, "Fibrillatory wave size in the diagnosis of heart disease," *Can. Med. Assoc. J.*, vol. 95, pp. 684–685, 1966.

[23] J. Morganroth, L. N. Horowitz, M. E. Josephson, and J. A. Kastor, "Relationship of atrial fibrillatory wave amplitude to left atrial size and etiology of heart disease. An old generalization re-examined," *Am. Heart J.*, vol. 97, pp. 184–186, 1979. DOI: 10.1016/0002-8703(79)90354-5

[24] M. H. Aysha and A. S. Hassan, "Diagnostic importance of fibrillatory wave amplitude: A clue to echocardiographic left atrial size and etiology of atrial fibrillation," *J. Electrocardiol.*, no. 21, pp. 247–251, 1988. DOI: 10.1016/0022-0736(88)90099-4

[25] A. Bollmann, K. Binias, K. Sonne, F. Grothues, H. Esperer, P. Nikutta, and H. Klein, "Electro-cardiographic characteristics in patients with nonrheumatic atrial fibrillation and their relation to echocardiographic parameters," *Pacing Clin. Electrophysiol.*, vol. 24, pp. 1507–1513, 2001. DOI: 10.1046/j.1460-9592.2001.01507.x

[26] J. L. Blackshear, R. E. Safford, and L. A. Pearce, "F-amplitude, left atrial appendage velocity, and thromboembolic risk in nonrheumatic atrial fibrillation. Stroke prevention in atrial fibrillation investigators," *Clin. Cardiol.*, vol. 19, pp. 309–313, 1996.

[27] A. C. Skanes, R. Mandapati, O. Berenfeld, J. M. Davidenko, and J. Jalife, "Spatiotemporal periodicity during atrial fibrillation in the isolated sheep heart," *Circulation*, vol. 98, pp. 1236–1248, 1998.

[28] J. Slocum, "Use of the autocorrelation function to detect atrial fibrillatory activity on the surface electrocardiogram," in *Proc. IEEE EMBS*, vol. 13, 1991.

[29] Z. Ihara, V. Jacquemet, J.-M. Vesin, and A. van Oosterom, "Adaptation of the standard 12-lead ECG system focusing on atrial electrical activity," in *Proc. Comput. Cardiol.*, vol. 32, pp. 203–206, IEEE Press, 2005.

[30] Z. Ihara, A. van Oosterom, V. Jacquemet, and R. Hoekema, "Adaptation of the 12-lead elec-trocardiogram system dedicated to the analysis of atrial fibrillation," *J. Electrocardiol.*, vol. 40, pp. 68.e1–68.e8, 2007. DOI: 10.1016/j.jelectrocard.2006.04.006

[31] M. Haissaguerre, P. Jais, D. C. Shah, A. Takahashi, M. Hocini, G. Quiniou, S. Garrigue, A. Le Mouroux, P. Le Metayer, and J. Clementy, "Spontaneous initiation of atrial fibrillation by ectopic beats originating in the pulmonary veins," *New Eng. J. Med.*, vol. 339, pp. 659–666, 1998. DOI: 10.1056/NEJM199809033391003

[32] S. Lazar, S. Dixit, F. E. Marchlinski, D. J. Callans, and E. P. Gerstenfeld, "Presence of left-to-right atrial frequency gradient in paroxysmal but not persistent atrial fibrillation in humans," *Circulation*, vol. 110, pp. 3181–3186, 2004. DOI: 10.1161/01.CIR.0000147279.91094.5E

[33] J. Slocum and K. Ropella, "Correspondence between the frequency domain characteristics of simultaneous surface and intra-atrial recordings of atrial fibrillation," in *Proc. Comput. Cardiol.*, pp. 781–784, IEEE Computer Society, 1994. DOI: 10.1109/CIC.1994.470069

[34] M. Holm, S. Pehrsson, M. Ingemansson, L. Sörnmo, R. Johansson, L. Sandhall, M. Sunemark, B. Smideberg, C. Olsson, and S. B. Olsson, "Non-invasive assessment of atrial refractoriness during atrial fibrillation in man—Introducing, validating, and illustrating a new ECG method," *Cardiovasc. Res.*, vol. 38, pp. 69–81, 1998. DOI: 10.1016/S0008-6363(97)00289-7

[35] S. Petrutiu, A. V. Sahakian, W. B. Fisher, and S. Swiryn, "Manifestation of left atrial events in the surface electrocardiogram," in *Proc. Comput. Cardiol.*, vol. 33, pp. 1–4, http://cinc.mit.edu, 2006.

[36] M. S. Hazen, T. H. Marwick, and D. A. Underwood, "Diagnostic accuracy of the resting electrocardiogram in detection and estimation of left atrial enlargement: an echocardiographic correlation in 551 patients," *Am. Heart J.*, vol. 122, pp. 823–828, 1991. DOI: 10.1016/0002-8703(91)90531-L

[37] W. A. MacLean, R. B. Karp, N. T. Kouchoukos, T. N. James, and A. L. Waldo, "P waves during ectopic atrial rhythms in man: A study utilizing atrial pacing with fixed electrodes," *Circulation*, vol. 52, pp. 426–434, 1975.

[38] F. G. Cosio, M. López-Gil, A. Goicolea, F. Arribas, and J. L. Barroso, "Radiofrequency ablation of the inferior vena cava-tricuspid valve isthmus in common atrial flutter," *Am. J. Cardiol.*, vol. 15, pp. 705–709, 1993. DOI: 10.1016/0002-9149(93)91014-9

[39] J. Ng, A. V. Sahakian, W. G. Fisher, and S. Swiryn, "Surface ECG vector characteristics of organized and disorganized atrial activity during atrial fibrillation," *J. Electrocardiol.*, vol. 37, pp. 91–97, 2004. DOI: 10.1016/j.jelectrocard.2004.08.031

[40] R. A. Warner, N. E. Hill, S. Mookherjee, and H. Smulyan, "Improved electrocardiographic criteria for the diagnosis of left anterior hemiblock," *Am. J. Cardiol.*, vol. 51, pp. 723–726, 1983. DOI: 10.1016/S0002-9149(83)80122-2

[41] J. Ng, A. V. Sahakian, W. G. Fisher, and S. Swiryn, "Atrial flutter loops derived from the surface ECG: Does the plane of the loop correspond anatomically to the macroreentrant circuit?," *J. Electrocardiol.*, vol. 36, pp. S181–186, 2003. DOI: 10.1016/j.jelectrocard.2003.09.055

[42] G. W. Botteron and J. M. Smith, "Quantitative assessment of the spatial organization of atrial fibrillation in the human heart," *Circulation*, vol. 93, pp. 513–518, 1996.

[43] K. M. Ropella, A. V. Sahakian, J. M. Baerman, and S. Swiryn, "The coherence spectrum. A quantitative discriminator of fibrillatory and nonfibrillatory cardiac rhythms," *Circulation*, vol. 80, pp. 112–119, 1989.

[44] H. J. Sih, D. P. Zipes, E. J. Berbari, and J. E. Olgin, "A high-temporal resolution algorithm for quantifying organization during atrial fibrillation," *IEEE Trans. Biomed. Eng.*, vol. 46, pp. 440–450, 1999. DOI: 10.1109/10.752941

[45] J. Jalife, O. Berenfeld, and M. Mansour, "Mother rotors and fibrillatory conduction: A mechanism of atrial fibrillation," *Cardiovasc. Res.*, vol. 54, pp. 204–216, 2002. DOI: 10.1016/S0008-6363(02)00223-7

CHAPTER 3

Atrial Activity Extraction from the ECG

Leif Sörnmo, Martin Stridh, and José Joaquín Rieta

3.1 INTRODUCTION

Since characterization of the atrial fibrillatory process often revolves around the determination of rate, a straightforward approach would be to perform spectral analysis on the ECG samples in intervals without ventricular activity, i.e., the TQ intervals [1]. However, such an approach does not fully exploit the ongoing nature of the fibrillatory process; samples during the QRST interval can be equally used for determining rate, provided that the ventricular activity has been first cancelled. The availability of additional samples reflecting atrial activity contributes not only to improved accuracy of the spectral estimate, but also circumvents the problem of vanishing TQ intervals at high heart rates.

The extraction of an atrial signal during AF requires nonlinear signal processing techniques since the atrial and ventricular activity overlap spectrally and therefore cannot be separated by linear filtering. Average beat subtraction (ABS) is the most widespread technique for atrial signal extraction and relies on the fact that atrial activity is uncoupled to ventricular activity during AF. Hence, subtraction of the average QRST complex produces a residual signal which is named the atrial signal. Since ABS is performed in individual leads, it becomes sensitive to alterations in the electrical axis, being manifested as large QRST-related residuals in the atrial signal. However, the effect of such alterations can be reduced using spatiotemporal QRST cancellation in which the average beats of adjacent leads are mathematically combined with the average beat of the analyzed lead in order to produce optimal cancellation (Section 3.2).

Another approach to atrial signal extraction exploits the property that the atrial and ventricular activities originate from different bioelectrical sources. The signals recorded on the body surface are assumed to be linear, instantaneous mixtures of the sources, but also of noise sources related to, e.g., muscular activity. Depending on the assumptions on the structure of the mixing matrix and the statistical properties of the sources, various techniques for source separation have been developed, requiring that multichannel recordings are available. Principal component analysis (PCA) and independent component analysis (ICA) are the two main representatives of such techniques which both have found their way into AF analysis (Sections 3.4 and 3.5, respectively). Under the assumptions that the signal sources are uncorrelated and the mixing matrix is orthogonal, PCA provides the optimal solution. However, ICA may be preferred in more general situations where the mixing matrix is assumed to have arbitrary structure and the sources are characterized by higher order statistics.

The above two approaches are the ones which dominate the literature, and, as such, they define the content of this chapter. Other methods have also been presented where ventricular activity is reduced through Wiener filtering using a time-delay artificial neural network [2], multiresolution analysis based on wavelet packets [3] and wavelets [4], Bayesian estimation [5], or empirical mode decomposition [6].

The issue of performance evaluation is particularly challenging when dealing with methods for atrial signal extraction because the accuracy of the resulting atrial signal cannot be easily quantified. As a consequence, authors have made use of indirect performance measures such as amplitude before and after cancellation and estimated atrial fibrillatory frequency. Performance may also be evaluated by means of simulated AF signals as these make it possible to judge how well the extracted atrial signal resembles the original one. Section 3.6 offers a brief overview of performance evaluation and related aspects such as the formulation of a simple AF simulation model.

3.2   AVERAGE BEAT SUBTRACTION AND VARIANTS

The ABS method was initially developed to facilitate the identification of P waves during ventricular tachycardia [7]. Later, this method was applied to AF analysis for the purpose of extracting the atrial fibrillatory waveforms, which are contiguous in nature [8, 9, 10, 11, 12, 13]. In both these applications, the ECG signal was processed on a single-lead basis so that an average beat, representative of the ventricular cycle, was subtracted from each individual heartbeat. The resulting residual signal would then, ideally, contain the fibrillatory waveforms subjected to further analysis.

The average beat is computed from an ensemble of sinus beats contained in the segment to be processed, assuming that the beat morphologies have already been labelled and grouped. Obviously, the larger the number of sinus beats available, the better the cancellation of atrial activity in the average beat. While the ensemble average is commonly used, other estimators may be preferable, for example, the median beat is more robust to spike artifacts. The exponential averager is yet another type of estimator that operates recursively rather than blockwise in order to track slow changes in QRST morphology [14, 15]. A comprehensive description of different types of estimators can be found in [16, Ch. 4].

Prior to subtraction, it is crucial to assure that the average beat and the QRST complex are well aligned in time to each other. If not, the resulting atrial signal will contain QRST residuals which render subsequent processing of the atrial signal considerably more difficult. Temporal alignment is required in any method that involves subtraction of the average beat. The alignment problem is defined by

$$\epsilon_{\min}^2 = \min_\tau \|\mathbf{x} - \mathbf{J}_\tau \bar{\mathbf{x}}\|^2 , \qquad (3.1)$$

where the vector $\mathbf{x}$ denotes N samples of the observed signal $x(n)$,

$$\mathbf{x} = \begin{bmatrix} x(0) \\ x(1) \\ \vdots \\ x(N-1) \end{bmatrix} . \qquad (3.2)$$

The average beat $\bar{\mathbf{x}}$ contains 2Δ additional samples so as to allow for temporal alignment of $\mathbf{x}$ relative $\bar{\mathbf{x}}$ using the shift matrix $\mathbf{J}_\tau$,

$$\mathbf{J}_\tau = \begin{bmatrix} \mathbf{0}_{N\times(\Delta+\tau)} & \mathbf{I}_{N\times N} & \mathbf{0}_{N\times(\Delta-\tau)} \end{bmatrix} , \qquad (3.3)$$

where τ denotes an integer time shift. The two matrices $\mathbf{0}$ and $\mathbf{I}$ denote the zero and identity matrix, respectively. The maximal alignment error that can be corrected is $\pm\Delta$. The minimization in (3.1) is performed as a grid search over all admissible values of τ, thus determining the N samples of $\bar{\mathbf{x}}$ which provides the best fit to $\mathbf{x}$. It should be noted that the minimization performed in (3.1) is equivalent to the more commonly used maximization of the crosscorrelation between $\mathbf{x}$ and $\bar{\mathbf{x}}$, i.e., the cross-term of the norm.

When ectopic beats are also present, it is necessary to create one beat average for each particular beat morphology before meaningful subtraction can be done. It is evident that the performance of the ABS method deteriorates as the number of ectopic beats per category decreases.

3.2.1 SPATIOTEMPORAL QRST CANCELLATION

Average beat subtraction relies on the assumption that an average beat represents each individual beat accurately. However, QRST morphology is often subject to minor changes due to variations in the orientation of the heart's electrical axis. Such variations are primarily due to respiratory activity: the electrical axis of the QRS complex varies as much as $10°$ during inspiration in the transversal plane and, accordingly, influence the precordial leads quite considerably [17]. Lead V2 is, in general, more sensitive to changes in position and orientation of the heart than the other precordial leads [18]. Due to the single-lead nature of ABS, such axis variations sometimes cause considerable QRST-related residuals. Since the variations occur on a beat-to-beat basis, the use of recursive beat-to-beat updating methods such as exponential averaging cannot handle this problem satisfactorily.

The spatiotemporal method [19] assumes that a multi-lead ECG is available, and, consequently, the data vector in (3.2) has to be replaced by a matrix $\mathbf{X}$ which contains N samples from L leads,

$$\mathbf{X} = \begin{bmatrix} \mathbf{x}_1 & \mathbf{x}_2 & \cdots & \mathbf{x}_L \end{bmatrix} . \qquad (3.4)$$

As before, the average beat $\tilde{\mathbf{X}}$ contains 2Δ additional samples in order to allow for temporal alignment of $\mathbf{X}$ relative $\tilde{\mathbf{X}}$ through the shift matrix $\mathbf{J}_\tau$.

Since atrial activity is assumed to be uncoupled to ventricular activity during AF, each observed beat $\mathbf{X}$ can be modeled as a sum of atrial activity $\mathbf{X}_A$, ventricular activity $\mathbf{X}_V$, and additive noise $\mathbf{W}'$,

$$\mathbf{X} = \mathbf{X}_A + \mathbf{X}_V + \mathbf{W}' . \tag{3.5}$$

The ventricular activity is modeled by

$$\mathbf{X}_V = \mathbf{J}_\tau \tilde{\mathbf{X}} \mathbf{S} , \tag{3.6}$$

where $\mathbf{S}$ is a spatial alignment matrix $(L \times L)$ which introduces the following properties in the cancellation process:

1. shifting of information between leads to compensate for variations in the electrical axis, and

2. scaling to compensate for variations in tissue conductivity and heart position which may affect the amplitude in different leads.

While $\mathbf{S}$ can have different structures, its definition as the product of a diagonal amplitude scaling matrix $\mathbf{D}$ and a rotation matrix $\mathbf{Q}$ was found to produce superior performance [19],

$$\mathbf{S} = \mathbf{D}\mathbf{Q}, \tag{3.7}$$

where the diagonal elements d_l of $\mathbf{D}$ are assumed to be positive valued.

Ultimately, the aim is to estimate the parameters $\mathbf{D}$, $\mathbf{Q}$, and τ from the observed signal $\mathbf{X}$, and then to subtract the resulting estimate of $\mathbf{X}_V$ from $\mathbf{X}$. However, combining (3.5) and (3.6), i.e.,

$$\mathbf{X} - \mathbf{J}_\tau \tilde{\mathbf{X}} \mathbf{S} = \mathbf{W}' + \mathbf{X}_A , \tag{3.8}$$

it is immediately clear that not only $\mathbf{W}'$ limits how well $\mathbf{J}_\tau \tilde{\mathbf{X}} \mathbf{S}$ will fit $\mathbf{X}$, but so will the atrial activity $\mathbf{X}_A$. In order to handle this problem, an intermediate estimate can be introduced, denoted $\tilde{\mathbf{X}}_A$, to be subtracted from $\mathbf{Y}$ prior to parameter estimation:

$$\mathbf{Y} - \mathbf{J}_\tau \tilde{\mathbf{X}} \mathbf{S} = \mathbf{W}' + \mathbf{X}_A - \tilde{\mathbf{X}}_A , \tag{3.9}$$

where

$$\mathbf{Y} = \mathbf{X} - \tilde{\mathbf{X}}_A . \tag{3.10}$$

The process of subtracting $\tilde{\mathbf{X}}_A$ in (3.9) is based on the availability of a so-called TQ-based fibrillation signal which, for example, can be computed as described below in Section 3.2.2.

The QRST cancellation parameters $\mathbf{D}$, $\mathbf{Q}$, and τ are estimated by solving the following minimization problem [19],

$$\epsilon^2_{min} = \min_{\mathbf{D}, \mathbf{Q}, \tau} \|\mathbf{Y} - \mathbf{J}_\tau \tilde{\mathbf{X}} \mathbf{D}\mathbf{Q}\|^2_F , \tag{3.11}$$

where ϵ^2 denotes the quadratic error defined by the Frobenius norm for an arbitrary matrix $\mathbf{A}$,

$$\|\mathbf{A}\|_F^2 = \text{tr}(\mathbf{A}\mathbf{A}^T) . \tag{3.12}$$

Unfortunately, minimization with respect to $\mathbf{Q}$ and $\mathbf{D}$ cannot be done independently of each other and a closed-form solution is difficult to find. Instead, an alternating, iterative approach can be employed in which the error, i.e., the Frobenius norm in (3.11), is minimized with respect to $\mathbf{Q}$, assuming that $\mathbf{D}$ is known. The minimization involves singular value decomposition (SVD) with which an arbitrary matrix $\mathbf{T}$ is decomposed into two orthonormal matrices, $\mathbf{U}$ and $\mathbf{V}$, and a diagonal matrix $\mathbf{\Sigma}$ that contains the singular values σ_l, i.e.,

$$\mathbf{T} = \mathbf{U}\mathbf{\Sigma}\mathbf{V}^T . \tag{3.13}$$

Setting $\mathbf{T} = \mathbf{D}^T\bar{\mathbf{X}}^T\mathbf{J}_\tau^T\mathbf{Y}$ and performing the SVD, it can be shown that the error is minimized for [20]

$$\hat{\mathbf{Q}} = \mathbf{U}\mathbf{V}^T . \tag{3.14}$$

Next, with an estimate of $\mathbf{Q}$ available, the diagonal entries of $\mathbf{D}$ can be estimated by [19]

$$\hat{d}_l = (\left[\mathbf{J}_\tau\bar{\mathbf{X}}\right]_l^T \left[\mathbf{J}_\tau\bar{\mathbf{X}}\right]_l)^{-1}(\left[\mathbf{J}_\tau\bar{\mathbf{X}}\right]_l^T \left[\mathbf{Z}\mathbf{Q}^{-1}\right]_l), \quad l = 1, \ldots, L , \tag{3.15}$$

where $[\cdot]_l$ denotes the lth column of the matrix. An improved estimate of $\mathbf{Q}$ can then be obtained from (3.14), using the estimate of $\mathbf{D}$, and so on.

Typically, a solution close to $\mathbf{Q} = \mathbf{D} = \mathbf{I}$ is desirable, and, therefore, the algorithm is initialized with $\mathbf{D}_0 = \mathbf{I}$. Obviously, the assumption inherent to the ABS method serves as the starting point for the spatiotemporal method. The rotation at step k, i.e., $\mathbf{Q}_k$, is then calculated from $\mathbf{D}_{k-1}$. Since

$$\|\mathbf{Y} - \mathbf{J}_\tau\bar{\mathbf{X}}\mathbf{D}_{k-1}\mathbf{Q}_k\|_F^2 \leq \|\mathbf{Y} - \mathbf{J}_\tau\bar{\mathbf{X}}\mathbf{D}_{k-1}\mathbf{Q}_{k-1}\|_F^2 \tag{3.16}$$

the error will be less or equal to that in the previous step. When $\mathbf{Q}_k$ is known, $\mathbf{D}_k$ can be calculated and, accordingly,

$$\|\mathbf{Y} - \mathbf{J}_\tau\bar{\mathbf{X}}\mathbf{D}_k\mathbf{Q}_k\|_F^2 \leq \|\mathbf{Y} - \mathbf{J}_\tau\bar{\mathbf{X}}\mathbf{D}_{k-1}\mathbf{Q}_k\|_F^2 . \tag{3.17}$$

This procedure is repeated until the difference in error between two successive iterations is sufficiently small. The algorithm will converge since minimization with respect to $\mathbf{Q}$ and $\mathbf{D}$ for each step, i.e., according to (3.16) and (3.17), will lower the error ϵ^2. In practice, convergence is typically achieved after 5 or 6 iterations.

Finally, minimization with respect to τ is solved through a grid search of τ in the interval $[-\Delta, \Delta]$, cf. the procedure defined in (3.1). Consequently, estimates of $\mathbf{Q}$ and $\mathbf{D}$ must be computed for all values of τ.

Figure 3.1 illustrates the performance of spatiotemporal cancellation and ABS, showing that the QRST-related residuals are considerably smaller for the former type of method.

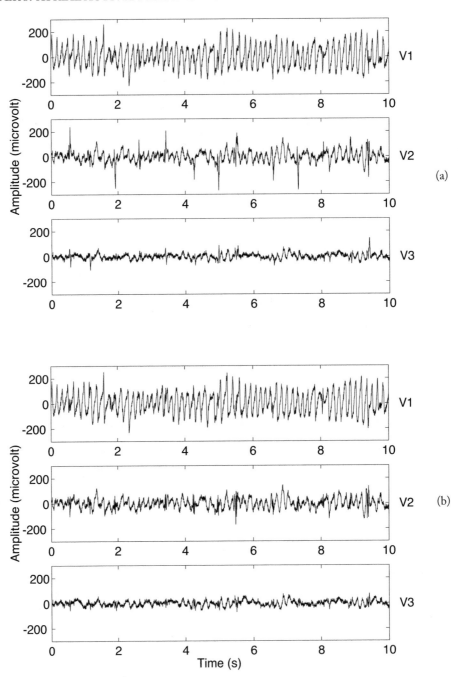

Figure 3.1: Three-lead atrial activity extraction using (a) average beat subtraction and (b) spatiotemporal cancellation.

3.2.2   TQ-BASED FIBRILLATION SIGNAL

The estimation of scaling and rotation parameters is hampered by the very presence of AF, thus implying that the resulting estimates are bound to be inaccurate. Therefore, in order to reduce the influence of AF, an intermediate AF signal, a TQ-based fibrillation signal, is reconstructed during this interval from the two enclosing TQ intervals, and subtracted from each beat prior to parameter estimation is performed. In [19], a simple approach is developed in which the fibrillatory cycles prior to the QRST complex is replicated during the QRST interval, but linearly weighted such that the weights decrease from one at the interval onset to zero at the end. An identical procedure is applied to the fibrillation cycles following the QRST complex, but in a time-reversed fashion. The TQ-based fibrillation signal is obtained by summing the two replicated waveforms. The cycle length to be replicated is estimated from the autocorrelation function of the two adjacent QT intervals. If only one of the enclosing intervals is long enough, only that interval is used for reconstructing the signal, whereas the signal is set to zero when both intervals are too short.

The introduction of a TQ-based fibrillation signal has been found to improve estimation accuracy quite substantially, while not much influencing the f-waves in the ECG. It should be noted that other approaches to deriving a TQ-based fibrillation signal can be considered as well. For example, the sinusoidal interpolation technique, described in Section 3.3 as part of the single-beat cancellation method, can be used in combination with spatiotemporal QRST cancellation.

3.2.3   SEPARATE QRS COMPLEX AND T WAVE CANCELLATION

The idea to process the QRS complex and the T wave with separate averages was initially suggested in [21], but later refined and evaluated in [22]. An important motivation to pursue separate processing is that the repolarization waveform changes considerably as a function of heart rate, whereas the depolarization waveform remains essentially unchanged. In this approach, the average beat $\bar{\mathbf{X}}$ is decomposed into two submatrices,

$$\bar{\mathbf{X}} = \begin{bmatrix} \bar{\mathbf{X}}_1 & \bar{\mathbf{X}}_2 \end{bmatrix} , \qquad (3.18)$$

so that $\bar{\mathbf{X}}_1$ and $\bar{\mathbf{X}}_2$ contain the averaged samples of the QRS interval and the JQ interval, respectively; the JQ interval starts at the J point and ends at the Q onset of the subsequent beat. In doing so, the two intervals can be processed differently with respect to

1. which beat intervals to include for beat averaging, and

2. how to fit the average beat interval to the observed signal.

The first processing consideration helps to produce more accurate averages as the inclusion criteria can be tailored to handle the respective properties of the depolarization and repolarization waveforms. It also implies that noise present in one interval does not have to influence handling of the other interval. The second consideration was translated to spatial optimization for the QRS interval, i.e., scaling and rotation, and temporal optimization over the JQ interval, i.e., performing ABS in this interval [22]. Using simulated ECG signals for performance evaluation, it was found that this approach produced a mean-square error (MSE) similar to that produced by spatiotemporal cancellation within the QRS interval. However, separate processing produced a lower MSE than did spatiotemporal cancellation within the T wave, thus suggesting that spatial optimization is less suitable for this interval.

A disadvantage with separate processing of the QRS and T intervals is that discontinuities may occur at the boundaries between the QRS complex and the T wave. This problem can to some extent be mitigated with lowpass filtering, for example, using a zero-phase, fifth-order Butterworth filter with cutoff frequency at 50 Hz [22].

3.3   SINGLE-BEAT CANCELLATION

The single-beat cancellation method processes one heartbeat at a time, and does not, in contrast to the previous methods described in this chapter, make use of information acquired from the beat averaging process [22]. The main idea is to estimate the dominant T wave morphology in each individual beat, using all available leads, and subtract it from the original ECG. The atrial activity within the QRS interval is estimated from interpolation of the atrial activity contained in the two enclosing JQ intervals, and, thus, no attempt is made to retrieve the atrial information being concealed within the QRS interval.

It has been observed that T wave morphology in different leads is quite similar—an observation which has given rise to the concept "the dominant T wave," found useful for representing individual T waves [23, 24]. The dominant T wave may be estimated from the $N \times L$ data matrix $\mathbf{X}$, containing the N samples of the JQ interval from the L-lead ECG, as the most significant eigenvector, i.e., computing the SVD of $\mathbf{X}$,

$$\mathbf{X} = \mathbf{U}\boldsymbol{\Sigma}\mathbf{V}^T , \tag{3.19}$$

and choosing that eigenvector (i.e., column) of $\mathbf{U}$ which corresponds to the largest singular value.

In order to provide flexible modeling of each individual T wave, the dominant T wave is defined as a linear combination of the most significant eigenvector and its time derivatives [22]. However, since the eigenvector contains noise which may influence the derivatives quite considerably, it was necessary to fit a smooth analytical function to the eigenvector and then compute the time derivatives from the fitted function. For this purpose, the following function has been employed,

$$f(t) = p_1 \left(p_2 + \frac{1}{1 + e^{p_3(t-p_5)}} \cdot \frac{1}{1 + e^{p_4(t-p_5)}} \right) , \tag{3.20}$$

where p_1 is a scale factor, p_2 defines the initial amplitude, p_3 and p_4 define the positive and negative slope, respectively, and p_5 defines the timing of the T wave apex. Due to the occasional presence of U waves, an additional function may be incorporated which accounts for such waves,

$$g(t) = p_6 e^{-(t-p_8)^2/p_7^2} , \tag{3.21}$$

where p_6 is a scale factor, p_7 defines the width, and p_8 is the timing of the U wave apex. Hence, the analytical function to be fitted to the most significant eigenvector is given by two terms,

$$f(t; \boldsymbol{\theta}_1) + g(t; \boldsymbol{\theta}_2) , \tag{3.22}$$

where the vectors $\boldsymbol{\theta}_1$ and $\boldsymbol{\theta}_2$ contain the parameters that define each respective function. A nonlinear optimization method (the Levenberg–Marquardt algorithm [25]) is employed to find estimates of the two vectors.

Next, the lth ECG lead is modeled as a linear combination of $f(t)$, two of its time derivatives, and $g(t)$,

$$x_l(t) = a_{0,l} f(t; \hat{\boldsymbol{\theta}}_1) + a_{1,l} f'(t; \hat{\boldsymbol{\theta}}_1) + a_{2,l} f''(t; \hat{\boldsymbol{\theta}}_1) + a_{3,l} g(t; \hat{\boldsymbol{\theta}}_2) , \tag{3.23}$$

where $a_{0,l}, \ldots, a_{3,l}$ denote the weight coefficients of the lth lead. These coefficients are obtained jointly for all leads using least squares estimation. Within the JQ interval, an atrial signal is then produced by subtracting the estimate of the dominant T wave, i.e., $\hat{x}_l(t)$, from the original ECG.

The atrial activity during the QRS interval is not extracted from the ECG, but is the result from interpolation of the atrial signals of the JQ intervals that enclose the QRS interval to be processed [22]. Employing sinusoidal interpolation, the coefficients that define the linear combination of sines and cosines at different frequencies are determined using least squares estimation. The frequencies are assumed to be uniformly distributed within the interval of atrial activity.

Finally, the atrial signal is obtained by concatenating the enclosing JQ intervals with the sinusoidal signal obtained from interpolation. Since the concatenated signal may contain jumps at the interval boundaries, lowpass filtering is performed as the last step of the single-beat cancellation method. The performance of the method is illustrated by Figure 3.2 which, in this case, is found to produce considerably smaller residuals than does ABS.

3.4   PRINCIPAL COMPONENT ANALYSIS

Principal components analysis performs an orthogonal linear transformation of the data such that the resulting principal components contain maximal information, measured by variance, and minimal redundancy, measured by correlation. The first principal component is a linear combination of the data that maximizes the joint variance, in the least squares sense. The second principal component holds the second largest variance, constrained to be uncorrelated with the first component, and so on.

The transformation is optimal in the sense that it retains the subspace that has the largest variance [26, 27]. Therefore, PCA is often used for dimensionality reduction of the data by keeping lower-order principal components, accounting for most of the variance of the data, while higher-order ones are ignored; however, as will be seen in this section, this is not always the case.

PCA-based extraction of atrial activity can be applied to single-lead ECGs, becoming a technique for finding data-dependent functions for cancellation of ventricular activity, or multi-lead ECGs, becoming a technique for blind source separation. Since these two techniques process the

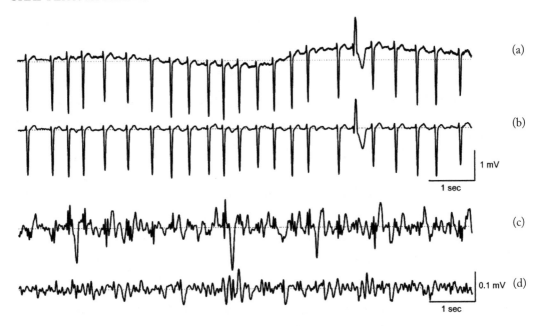

Figure 3.2: (a) 10-s ECG from lead V2. (b) Baseline corrected ECG. (c) Atrial signal extracted with average-beat subtraction. (d) Atrial signal extracted with the single-beat method. (Reprinted from [22] with permission.)

ECG signal in different ways, relying on different definitions of the data matrix $\mathbf{X}$, the techniques are described in separate sections below.

Principal component analysis is based on the assumption that the signal vector $\mathbf{x}$ is a zero-mean random process, characterized by the correlation matrix $\mathbf{R}_x = E[\mathbf{x}\mathbf{x}^T]$. The principal components $\mathbf{w} = \begin{bmatrix} w_1 & w_2 & \cdots & w_N \end{bmatrix}^T$ result from an orthogonal linear transformation $\mathbf{\Psi} = \begin{bmatrix} \boldsymbol{\psi}_1 & \boldsymbol{\psi}_2 & \cdots & \boldsymbol{\psi}_N \end{bmatrix}$ of $\mathbf{x}$,

$$\mathbf{w} = \mathbf{\Psi}^T \mathbf{x} , \tag{3.24}$$

which rotates $\mathbf{x}$ so that the elements of $\mathbf{w}$ become mutually uncorrelated. In order to obtain the set of N principal components, the eigenvector equation for $\mathbf{R}_x$ needs to be solved [26],

$$\mathbf{R}_x \mathbf{\Psi} = \mathbf{\Psi} \mathbf{\Lambda} . \tag{3.25}$$

Here, $\mathbf{\Lambda}$ is a diagonal matrix with the eigenvalues $\lambda_1, \lambda_2, \ldots, \lambda_N$, each eigenvalue reflecting the variance of the corresponding principal component. The eigenvector $\boldsymbol{\psi}_1$ (also known as the "principal component direction"), corresponding to the largest eigenvalue, is associated with the largest variance of the signal, and so on. Since $\mathbf{R}_x$ is rarely known in practice, it has to be replaced in (3.25) by the sample correlation matrix $\hat{\mathbf{R}}_x$, being estimated from $\mathbf{X}$.

3.4.1   SINGLE-LEAD ANALYSIS

In single-lead PCA, the ventricular activity to be cancelled is characterized in terms of intrabeat correlation, and, consequently, temporal properties are exploited rather than spatial ones [28]. The data matrix contains an ensemble of M single-lead beats with N samples, represented by the $N \times M$ data matrix

$$\mathbf{X} = \begin{bmatrix} \mathbf{x}_1 & \mathbf{x}_2 & \cdots & \mathbf{x}_M \end{bmatrix} . \tag{3.26}$$

It should be noted that this definition of $\mathbf{X}$ differs from the one introduced in (3.4) which contains several leads of one single beat. The number of beats M should be chosen large enough to produce a useful estimate of $\mathbf{R}_x$. While $\mathbf{X}$ may be allowed to contain beats of different morphologies, it may be desirable to select beats with similar morphology as such a choice implies that a smaller value of M is required. The N samples from each beat are selected with reference to a fiducial point such that the entire QRST complex is included in the segment, possibly overlapping with adjacent beats. The $N \times N$ sample correlation matrix is obtained by

$$\hat{\mathbf{R}}_x = \frac{1}{M} \mathbf{X} \mathbf{X}^T , \tag{3.27}$$

and used in (3.25) for eigenvector/eigenvalue computation.

Applying PCA to the ensemble of beats, the associated pattern of principal components reflects the degree of morphologic beat-to-beat variability: When the first principal component is much larger than the other components, the ensemble of QRST waveforms exhibits low morphologic variability, whereas a slow fall-off of the principal component values indicates large variability. In terms of resulting eigenvalues and eigenvectors, the following observations have been done [28]:

1. The eigenvector corresponding to the largest eigenvalue is related to the dominant QRST morphology since the ventricular activity exhibits the largest variance in the ECG.

2. The next few eigenvectors correspond to the dynamics of the QRST waveform. In case of a stable QRST morphology, these components are usually missing.

3. Next, there is a number of eigenvectors related to atrial activity.

4. The remaining eigenvectors correspond to noise of various origin such as muscular activity and electrode motion artifacts.

Since PCA projects each beat on the "ventricular subspace," a QRST waveform can be estimated and removed by subtracting a linear combination of the "ventricular" eigenvectors. Alternatively, atrial activity can be extracted from each beat by considering the projection on the "atrial subspace" (Figure 3.3). Needless to say, the performance of this technique is critically dependent on the algorithm being employed for subspace identification; poor identification leads to that the extracted atrial activity contains unwanted components. The identification aspect is discussed in [28], however, the details on the algorithm employed for such identification were not mentioned. It was found that the dimension of the atrial subspace ranges from 4–10, depending on the patient analyzed. One possible approach to automatic identification of the atrial subspace is to perform spectral analysis of all components and select those components which exhibit a dominant spectral peak in the 3–12 Hz interval; see Chapter 4 for information on the spectral properties of AF.

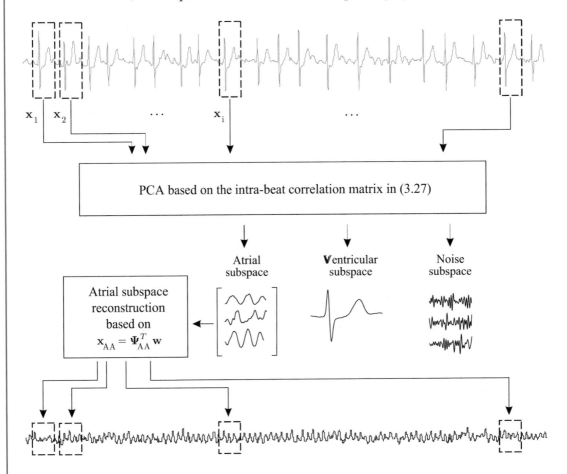

Figure 3.3: Block diagram of single-lead PCA to estimate atrial activity during AF. The correlation matrix is computed from the segmented ECG signal (top). The reconstruction of the atrial signal (bottom) is obtained by using the eigenvectors that define the atrial subspace $\mathbf{\Psi}_{AA}$ and concatenation. (Reprinted from [28] with permission.)

Single-lead PCA extraction of atrial activity may be considered as a generalization of the above-mentioned ABS since a template beat for subtraction is created as a linear combination of eigenvectors. This template beat is, however, likely to provide a better fit to the QRST waveforms than does the average beat [28].

3.4.2 MULTI-LEAD ANALYSIS

Another approach to the extraction of atrial activity is to explore the redundant information of multi-lead ECGs [29, 30]. By applying PCA, it is possible to decompose the multi-lead ECG so that the most representative component is the one which corresponds to ventricular activity, whereas the next few components correspond to variability in ventricular activity (cf. the single-lead case above). Among the next principal components it is usually possible to find a signal which contains atrial activity.

In multi-lead PCA, the definition of the data matrix is identical to that used for spatiotemporal QRST cancellation given in (3.4), i.e.,

$$\mathbf{X} = \begin{bmatrix} \mathbf{x}_1 & \mathbf{x}_2 & \cdots & \mathbf{x}_L \end{bmatrix} , \tag{3.28}$$

where L denotes the number of leads, which, in most studies, equals eight because the standard 12-lead ECG is analyzed (four of the leads are redundant). In contrast to single-lead analysis, segmentation of the data with reference to the fiducial point is not needed here.

Rather than characterizing the intrabeat correlation as in (3.27), multi-lead PCA exploits the interlead correlation by computing the $L \times L$ sample correlation matrix,

$$\hat{\mathbf{R}}_x = \frac{1}{N}\mathbf{X}^T\mathbf{X} . \tag{3.29}$$

As before, the eigenvectors $\mathbf{\Psi}$ that result from diagonalization of $\hat{\mathbf{R}}_x$ define the orthogonal linear transformation used to compute the principal components. In this case, the principal components are computed for each sample n by

$$\mathbf{w}(n) = \mathbf{\Psi}^T\mathbf{x}(n) , \tag{3.30}$$

where

$$\mathbf{x}(n) = \begin{bmatrix} x_1(n) \\ x_2(n) \\ \vdots \\ x_L(n) \end{bmatrix} . \tag{3.31}$$

Figure 3.4 presents an example where PCA is applied to an ECG with AF. The atrial activity can be identified as the fourth principal component, whereas the three first components contain ventricular activity (as well as some noise). The fifth and higher-order components contain noise which largely is of muscular origin. Similar to single-lead PCA, algorithms for automatic identification of the atrial subspace remain to be devised.

3.5 INDEPENDENT COMPONENT ANALYSIS

3.5.1 MODEL AND METHODS

Another useful approach to the extraction of atrial activity from the surface ECG is to assume that the observed signal is a mixture of different signal sources of atrial, ventricular, and extracardiac

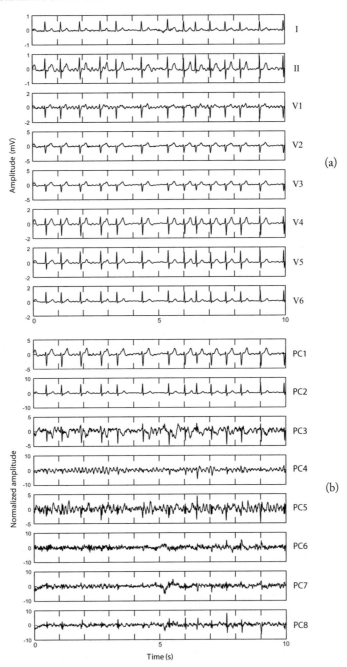

Figure 3.4: Example of atrial signal extraction during AF using a multi-lead PCA approach. (a) The original eight ECG leads and (b) the corresponding principal components.

origin. Although blind separation of signal sources can be accomplished by the above-mentioned second-order PCA approach, it is often desirable to develop a technique which exploits higher order statistics in order to achieve better performance. The following linear model is the starting point for such separation:

$$\mathbf{x}(n) = \mathbf{A}\mathbf{s}(n) , \tag{3.32}$$

where $\mathbf{A}$ denotes an $L \times L$ instantaneous mixing matrix, $\mathbf{s}(n)$ a vector with L different source signals at time n,

$$\mathbf{s}(n) = \begin{bmatrix} s_1(n) \\ s_2(n) \\ \vdots \\ s_L(n) \end{bmatrix} , \tag{3.33}$$

and $\mathbf{x}(n)$ another vector with the observations of L different "sensor" signals obtained from the ECG electrodes. For the situation when the number of sensors and sources differs, it is commonly assumed that there are at least as many sensors as there are sources.

A fundamental assumption is that the different sources $s_1(n)$, $s_2(n)$, $\ldots$, $s_L(n)$ are statistically mutually independent at each time instant n,

$$p(\mathbf{s}(n)) = \prod_{l=1}^{L} p_l(s_l(n)) , \tag{3.34}$$

where each source is characterized by its individual non-Gaussian probability density functions $p_l(s_l(n))$. It is obvious that since neither the instantaneous mixing matrix $\mathbf{A}$ is available, nor can the signal sources $\mathbf{s}(n)$ be observed, the term "blind source separation" is well motivated [31].

Independent component analysis is a powerful technique for blind source separation whose purpose is to find a linear transformation,

$$\mathbf{y}(n) = \mathbf{B}\mathbf{x}(n) , \tag{3.35}$$

so that the resulting components of $\mathbf{y}(n)$ become independent in the statistical sense, serving as estimates of the source signals $\mathbf{s}(n)$. Since strict statistical independence cannot be achieved in practice, the matrix $\mathbf{B}$ is chosen so that some suitable function designed to measure independence is maximized, usually being synonymous to maximizing non-Gaussianity. The linear transformation in (3.35) is, in contrast to PCA, no longer constrained to be orthogonal, but can possess any structure as long as $\mathbf{B}$ has full column rank.

A description of methods for determining the independent components $\mathbf{y}(n)$ is far beyond the scope of the present text. Here, we will just mention two concepts which are central to ICA, namely, 1. a quantitative measure of non-Gaussianity, and 2. its maximization. The interested reader is referred to the comprehensive literature available on this topic, see, e.g., [32]. The property of non-Gaussian independent components is crucial to the estimation of the ICA model, and, therefore, a measure for quantifying the degree of non-Gaussianity is required. Kurtosis constitutes the classic measure as it quantifies the relative peakedness of a distribution with respect to a Gaussian distribution. Kurtosis is, for a scalar random variable y with unit variance, defined by

$$\text{kurt}[y] = E\left[y^4\right] - 3 . \tag{3.36}$$

The kurtosis is zero for a Gaussian random variable, whereas it is nonzero for virtually all non-Gaussian random variables; it is negative for sub-Gaussian ("spiky") probability density functions, and positive for super-Gaussian ("flat") probability density functions. The absolute value of the kurtosis is used as a measure of non-Gaussianity. It can be shown that other measures of non-Gaussianity such as negentropy and mutual information are related to kurtosis [32].

The other central concept in ICA is the maximization of the chosen non-Gaussianity measure, done with respect to the elements of the demixing matrix $\mathbf{B}$, so as to produce the independent components. For example, the first element of the first independent component, i.e., $y_1(n)$, is a linear combination of the observations $x_1(n), x_2(n), \ldots, x_L(n)$, weighted with the elements of the first row in $\mathbf{B}$. These weights are chosen so that the absolute value of the kurtosis of $y_1(n)$ is maximized, employing a gradient-based method or some other suitable optimization method (of which the so-called FastICA algorithm is the most popular). The weights of the remaining rows may be determined by performing repeated row-wise optimization [32].

The amplitude of the resulting independent components $\mathbf{y}(n)$ cannot be determined since both $\mathbf{s}(n)$ and $\mathbf{A}(n)$, defining the observed signal $\mathbf{x}(n)$, are unknown quantities. As a result, the components are often normalized to have unit variance. Furthermore, it is not possible to determine the order of the independent components, implying that one has to potentially treat each component as the one being of interest. In contrast to PCA, where variance ordering is implicit, some other technique has to be used with ICA in order to identify the components of interest for further analysis.

It should be noted that the observed signal is typically preprocessed before the independent components are determined. Preprocessing includes centering, i.e., subtraction of the mean, and whitening, i.e., decorrelation and variance normalization, so as to facilitate the above-mentioned maximization of non-Gaussianity. Whitening, which may be accomplished with PCA, reduces dimensionality and increases the performance of the maximization method quite considerably.

3.5.2   APPLICATION TO ECG SIGNALS

Independent component analysis has been investigated in a number papers for the purpose of extracting atrial activity [33, 34, 35, 36]; see also [37] for the use of ICA in other ECG applications. This technique is able to extract atrial activity even in ECG signals as short as just a few seconds. This property stands in contrast to ABS and its variants where several beats are required to produce a usable average. Another property of ICA is that the independent component with atrial activity needs to be identified with some suitable technique before further AF-specific processing can be done; see below. The selected component can be viewed as a "global" atrial signal which contains contributions from all leads and whose amplitude does not easily translate to clinical terms (the component variance is usually normalized). As a result, lead-related information is lost as the independent components derive from one or several signal sources and not from a particular position on the body surface.

Model Assumptions

Successful application of ICA to atrial activity extraction requires that the above-mentioned assumptions on signal properties are approximately valid. Obviously, the assumptions cannot be easily corroborated since the source signals cannot be observed, however, various arguments have been put forward which suggest that ICA is suitable for this application.

The assumption of *statistical independence* involves the sources of ventricular and atrial origin as well as extracardiac noise sources such as muscular activity, respiration, and electrode movement. While the two cardiac activities are strongly coupled in the normal heart, they can be treated as independent statistical processes during AF as the atrial wavefronts lead to ventricular depolarization at highly irregular time instants [33]; see also Chapter 5. As was pointed out in Section 3.2, this assumption is also crucial for methods which are based on ABS. The assumption of independence between cardiac and extracardiac sources seems to be largely acceptable, although respiratory noise and ventricular activity are dependent since respiration to a certain degree modulates QRS amplitude.

The assumption of *non-Gaussian source signals* is valid when it comes to ventricular activity, because histogram analysis of the amplitude of ECGs at high signal-to-noise ratios (SNRs) shows that the kurtosis is much larger than zero, i.e., the ventricular activity is clearly super-Gaussian [33, 37]. On the other hand, this assumption is less valid for atrial activity as its kurtosis may approach zero [35], although it was initially suggested to have a sub-Gaussian distribution [33]. The statistical distribution of other extracardiac sources has not been given much attention in the literature. While certain types of noise are clearly super-Gaussian, e.g., powerline interference, other types such as muscular activity are approximately Gaussian [37], implying that the independent component(s) with atrial activity will contain a certain amount of muscular noise.

The validity of the assumption of *linear, instantaneous mixing*, described by the matrix $\mathbf{A}$, has been motivated by the structure of the solution to the so-called forward problem [33]; see also [38]. In that solution, the electrical potential on the body surface is obtained by adding the partial contributions of the potentials on the epicardial surface, each point being weighted by a linear, instantaneous transfer coefficient. The coefficients account for the conductivity of the human torso when approximated as an isotropic, homogeneous volume conductor. It should be noted, however, that the validity of this assumption may be questioned because the cardiac source rotates over time due to, e.g., respiration [37]. This complication calls for more advanced ICA methods which have the capability to track a time-varying mixture $\mathbf{A}$.

The assumption of a convolutive mixture instead of an instantaneous mixture has been investigated for atrial activity extraction [39]. In this model, which obviously is more complex, each element of $\mathbf{A}$ is a linear, time-invariant filter instead of a scalar, and, consequently, memory is introduced. The results showed, however, that the linear instantaneous mixture model yields better performance than do methods based on convolutive ICA using finite impulse response filters.

Atrial Component Identification

A crucial step in ICA-based atrial activity extraction is to identify the independent component(s) which contains atrial activity. The first algorithm proposed for this purpose made use of kurtosis-based reordering of the components, relying on the assumption that sub-Gaussian sources are associated with atrial activity, approximately Gaussian ones with various types of noise and artifacts, whereas super-Gaussian sources are associated with ventricular activity [33]. The following estimator of kurtosis was used,

$$\kappa_i = \frac{1}{N} \sum_{n=0}^{N-1} \left(\frac{y_i(n) - \hat{\mu}}{\hat{\sigma}} \right)^4 - 3 , \qquad (3.37)$$

where $y_i(n)$ denotes the ith independent component with N samples, $\hat{\mu}$ the sample mean, and $\hat{\sigma}$ the sample variance. Figure 3.5 illustrates the outcome of ICA for a 12-lead ECG recording, the independent components being displayed in increasing order of kurtosis. The atrial activity is mostly contained in the first independent component. The corresponding amplitude histograms for the first and the last components are presented in Figure 3.6. In this particular case, the "atrial" component has sub-Gaussian character, whereas the "ventricular" has super-Gaussian as indicated by the kurtosis.

Since information on kurtosis alone is insufficient for accurate identification of the atrial component, kurtosis reordering was combined with power spectral analysis of the sub-Gaussian components to detect when a dominant spectral peak, reflecting atrial rate, was present or not. It is commonly accepted that atrial rate is reflected by a peak whose frequency confined to the interval 3–12 Hz; see Chapter 4.

Another approach to atrial component identification was later presented in [35], where kurtosis reordering and spectral analysis are supplemented with another technique with which ventricular components are excluded from further processing and only components with possible, atrial activity are retained. Since the kurtosis of the ventricular components is usually very high, they can be excluded with a simple threshold test. It was found that a threshold of about 1.5 retained components with atrial activity, but excluded components with QRS complexes. The nonventricular components, i.e., atrial activity, noise, and artifacts, with kurtosis close to zero, are separated using second-order blind identification (SOBI). This technique aims at separating a mixture of uncorrelated sources with different spectral content through second-order statistical analysis which also takes into consideration the source temporal information [40]. For this purpose, SOBI seeks a transformation that simultaneously diagonalizes several correlation matrices at different lags. Since, in general, no transformation may exist that accomplish such a strict condition, a function is instead employed that objectively measures the degree of joint (approximate) diagonalization at different lags. Similar to the ICA-based technique, the atrial component resulting from ICA–SOBI is selected as the one which has a spectral peak in the interval 3–12 Hz.

The ICA–SOBI technique for atrial activity extraction is schematically presented in Figure 3.7, and its performance is illustrated in Figure 3.8 in relation to the ICA-based technique. This example

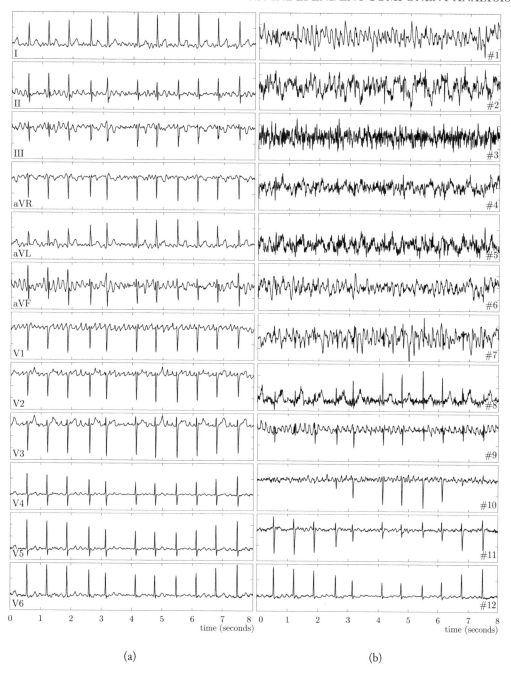

Figure 3.5: Example of atrial activity extraction using ICA. (a) The original 12-lead ECG and (b) the associated independent components. (Reprinted from [33] with permission.)

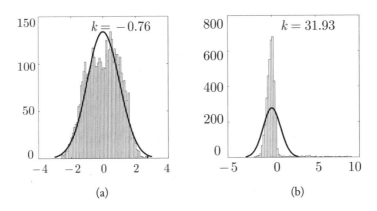

(a)   (b)

Figure 3.6: Amplitude histogram of two independent components displayed in Figure 3.5, namely: (a) component #1 with atrial activity and (b) component #12 with ventricular activity. The Gaussian distribution is superimposed for comparison. The kurtosis value for each histogram is indicated. (Reprinted from [33] with permission.)

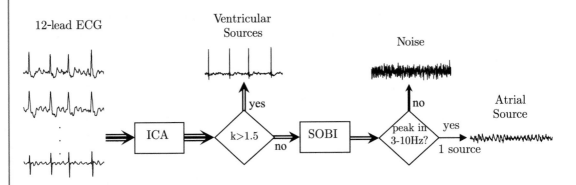

Figure 3.7: Block diagram of the ICA–SOBI technique for atrial activity extraction. Components whose kurtosis exceeds 1.5 are excluded. (Reprinted from [35] with permission.)

suggests that the spectral peak of the atrial component is more concentrated when employing ICA–SOBI, although the spectral peak can be discerned in both power spectra.

Finally, it should be noted that the ICA method has been modified to also include *a priori* information on the kurtosis sign of the source signals [36]. In doing so, the method combines spatial and temporal information without becoming computationally demanding.

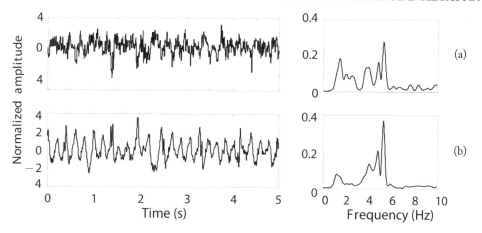

Figure 3.8: Atrial activity extraction using (a) ICA and (b) ICA–SOBI. The atrial signal and its corresponding power spectrum are displayed in the left and right diagram, respectively. (Adapted from [35].)

3.6   PERFORMANCE EVALUATION

The evaluation of QRST cancellation performance constitutes a challenging problem as the quality of the resulting atrial signal cannot be easily translated to parametric terms. This problem has been addressed using either simulated ECG signals with well-defined AF characteristics (Section 3.6.1) or by defining performance measures which capture certain critical amplitude or frequency characteristics of the atrial signal (Section 3.6.2). While none of these approaches to performance evaluation have been accepted as common standard, they are both considered in the literature.

3.6.1   SIMULATION MODELS

AF replication model.   The generation of an AF signal can be accomplished using the atrial fibrillatory ECG itself as the starting point [35]. Since TQ intervals contain f-waves free of ventricular activity, interpolation between two successive TQ intervals can be used to fill in the intermediate QT interval with atrial activity. Identical to the generation of the TQ-based fibrillation signal, described earlier in Section 3.2.2, the fibrillatory waves preceding the QT interval are replicated within this interval with linear weighting; the waves following the QT interval are replicated in the same way but time-reversed. The simulated atrial signal results from summation of the two replicated and weighted signals.

Since the AF replication model was developed for performance evaluation of ICA-based methods, it was argued that ventricular activity should be derived from the same patient as atrial activity; if not, the mixing matrix of atrial activity would generally be different from that associated with ventricular activity [35]. In order to derive both these activities from the same patient, patients who underwent cardioversion were studied. The atrial activity was extracted prior to cardioversion and, following successful cardioversion, the ventricular activity was extracted from the normal sinus rhythm, though with the P waves removed. The simulated ECG signal was obtained by summation of the atrial and ventricular activity.

The replication model produces realistic ECG signals as illustrated in Figure 3.9. However, various signal properties such as AF frequency and waveform morphology are not easily modified for testing purposes with this model. It should be noted that the AF replication model cannot be used at high heart rates as the TQ interval will eventually vanish.

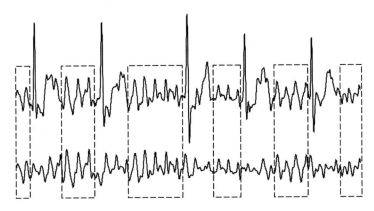

Figure 3.9: The generation of an atrial signal using the AF replication model is based on the samples in successive TQ intervals (indicated with boxes). (Reprinted from [35] with permission.)

Sawtooth model. This model is a simple mathematical tool for producing a signal which resembles the spectral pattern of atrial fibrillatory waves of the ECG signal [19, 41, 42]. The sawtooth signal is the linear combination of a sinusoid and its $M - 1$ harmonics. It can be made nonstationary by assigning time-varying properties to the fundamental frequency as well as to the amplitude of the sinusoid. In each lead, the atrial activity is modeled by

$$x(n) = \sum_{m=1}^{M} a_m(n) \sin\left(m \cdot \omega_0 n + \frac{\Delta f}{f_f} \sin(\omega_f n)\right), \quad n = 0, \ldots, N - 1, \qquad (3.38)$$

where the fundamental frequency $\omega_0 = 2\pi f_0$ has the maximum frequency deviation Δf and the modulation frequency $\omega_f = 2\pi f_f$. The time-varying amplitude of the simulated fibrillatory wave-forms is defined so that the resulting signal has sawtooth characteristic, using

$$a_m(n) = \frac{2}{m\pi}\left(a_m + \Delta a_m \sin(\omega_a n)\right), \qquad (3.39)$$

where a_m denotes the sawtooth amplitude, Δa_m the modulation amplitude, and $\omega = 2\pi f_a$ the amplitude modulation frequency.

A test signal suitable for performance evaluation can be generated by adding the simulated AF signal $x(n)$ to an ECG signal recorded from healthy subjects with normal sinus rhythm, but with the P waves first removed. The main advantage of taking the ventricular activity from healthy subjects is that the inherent variations in QRS morphology, e.g., due to respiration, are present in the simulated signal. The fact that the RR interval pattern during AF is highly irregular, while regular in the simulated ECG signal, is uncritical as QRST cancellation methods do not rely on the RR interval pattern anyhow.

Figure 3.10 illustrates the appearance of a three-lead simulated signal using the above sawtooth model such that each lead results from repeated use of (3.38), but with lead-dependent parameter values. In this example, the simulated atrial fibrillatory pattern has relatively low fibrillation frequency (6 Hz), five harmonics (i.e., $M = 6$), and a large AF amplitude, especially in lead V1.

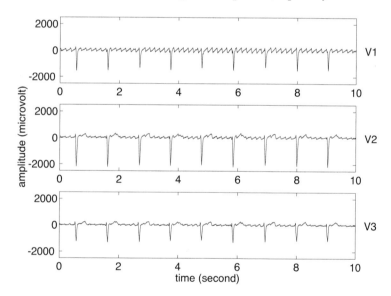

Figure 3.10: The generation of a three-lead ECG signal using the sawtooth model of atrial activity.

It is obvious that the sawtooth model is purely phenomenological in nature and does not account for any physiological insights on AF. Still, this model can be useful for technical evaluation of QRST cancellation performance as well as time–frequency analysis of AF because of its flexibility to define the fibrillation frequency and related variability.

Biophysical model.   QRST cancellation performance has also been evaluated using a much more sophisticated model which is capable of simulating surface ECGs as well as electrograms [6]. This model makes use of electroanatomical information on the atria as well as volume conduction theory. A more detailed description of the biophysical model is found in Chapter 7; see also [43].

3.6.2   PERFORMANCE MEASURES

Clearly, all of the above simulation models have in common that the desired AF signal is completely known. Therefore, it is possible to compute various performance measures between the true and the estimated AF signals such as the root mean-square error and the crosscorrelation coefficient. The latter measure reflects similarity in f-wave morphology but not differences in amplitude between the two signals.

Although it is difficult to quantify QRST cancellation performance in atrial signals extracted from real ECGs, a number of performance measures has nonetheless been suggested in the literature which exploit certain time or frequency domain properties. Such measures are indirect in nature and can thus only offer a relatively crude description of performance.

Time domain.   The characteristics of f-wave amplitude have been used for the purpose of studying cancellation performance [44]. The suppression of ventricular activity can be quantified by comparing the mean value of the peak-to-peak amplitudes in different QRST intervals, computed before and after QRST cancellation. Another performance measure quantifies amplitude changes in TQ intervals by comparing the mean peak-to-peak amplitude in these intervals, computed before and after QRST cancellation. Differences in amplitude should be as small as possible since the cancellation method should not modify the f-wave amplitude.

Frequency domain.   Another approach to define a performance measure is to quantify certain spectral properties of the estimated AF signal. The so-called *spectral concentration* is one such measure, being defined as the ratio between the power in a small interval centered around the dominant spectral peak and the total spectral power [35, 39]; the onset and end of this interval has been set equal to 0.82 and 1.17 of the peak frequency, respectively. The rationale for using this performance measure is that large QRST residuals would cause the peak to be less concentrated, and vice versa. Consequently, it may seem desirable to attain spectral concentration which is as large as possible.

3.7   CONCLUSIONS

The choice of approach for atrial activity extraction can be guided by the decidedly different modes by which the ECG signal is processed. Using ABS or its variants, the atrial signal is extracted in specific leads with the aim not to modify f-wave morphology. On the other hand, both PCA and ICA derive a global atrial signal with contributions from all leads, though dominated by the lead containing the largest atrial activity. A recording length of at least 10 s is required for adequate computation of an average beat, whereas the recording length can be shorter when PCA or ICA is used. The variants of ABS can process recordings with any number of leads, whereas PCA and ICA exploit spatial diversity and thus require recordings with a certain minimum number of leads, preferably the 12-lead ECG.

Bibliography

[1] D. S. Rosenbaum and R. J. Cohen, "Frequency based measures of atrial fibrillation in man," in *Proc. IEEE EMBS*, vol. 12, pp. 582–583, 1990.

[2] C. Vásquez, A. Hernández, F. Mora, G. Carrault, and G. Passariello, "Atrial activity enhancement by Wiener filtering using an artificial neural network," *IEEE Trans. Biomed. Eng.*, vol. 48, pp. 940–944, 2001. DOI: 10.1109/10.936371

[3] C. Sánchez, J. Millet, J. J. Rieta, F. Castells, J. Ródenas, R. Ruiz-Granell, and V. Ruiz, "Packet wavelet decomposition: An approach for atrial activity extraction," in *Proc. Comput. Cardiol.*, pp. 33–36, IEEE Press, 2002.

[4] C. Sánchez, J. J. Rieta, F. Castells, R. Alcaraz, and J. Millet, "Wavelet blind separation: A new methodology for the analysis of atrial fibrillation from Holter recordings," in *Proc. Comput. Cardiol.*, pp. 417–420, IEEE Press, 2004. DOI: 10.1109/CIC.2004.1442962

[5] O. Meste and N. Serfaty, "QRST cancellation using Bayesian estimation for the auricular fibrillation analysis," in *Proc. IEEE EMBS*, pp. 7083–7086, 2005. DOI: 10.1109/IEMBS.2005.1616138

[6] M. Lemay and J.-M. Vesin, "QRST cancellation based on the empirical mode decomposition," in *Proc. Comput. Cardiol.*, vol. 33, pp. 561–564, http://cinc.mit.edu, 2006.

[7] J. Slocum, E. Byrom, L. McCarthy, A. V. Sahakian, and S. Swiryn, "Computer detection of atrioventricular dissociation from surface electrocardiograms during wide QRS complex tachycardia," *Circulation*, vol. 72, pp. 1028–1036, 1985.

[8] J. Slocum, A. V. Sahakian, and S. Swiryn, "Diagnosis of atrial fibrillation from surface electrocardiograms based on computer-detected atrial activity," *J. Electrocardiol.*, vol. 25, pp. 1–8, 1992. DOI: 10.1016/0022-0736(92)90123-H

[9] S. Shkurovich, A. V. Sahakian, and S. Swiryn, "Detection of atrial activity from high-voltage leads of implantable ventricular defibrillators using a cancellation technique," *IEEE Trans. Biomed. Eng.*, vol. 45, pp. 229–234, 1998. DOI: 10.1109/10.661270

[10] M. Holm, S. Pehrsson, M. Ingemansson, L. Sörnmo, R. Johansson, L. Sandhall, M. Sunemark, B. Smideberg, C. Olsson, and S. B. Olsson, "Non-invasive assessment of atrial refractoriness during atrial fibrillation in man—Introducing, validating, and illustrating a new ECG method," *Cardiovasc. Res.*, vol. 38, pp. 69–81, 1998. DOI: 10.1016/S0008-6363(97)00289-7

[11] A. Bollmann, N. Kanuru, K. McTeague, P. Walter, D. B. DeLurgio, and J. Langberg, "Frequency analysis of human atrial fibrillation using the surface electrocardiogram and its response to ibu-tilide," *Am. J. Cardiol.*, vol. 81, pp. 1439–1445, 1998. DOI: 10.1016/S0002-9149(98)00210-0

[12] Q. Xi, A. V. Sahakian, and S. Swiryn, "The effect of QRS cancellation on atrial fibrillatory wave signal characteristics in the surface electrocardiogram," *J. Electrocardiol.*, vol. 36, pp. 243–249, 2003. DOI: 10.1016/S0022-0736(03)00046-3

[13] F. Beckers, W. Anne, B. Verheyden, C. van der Dussen de Kestergat, E. van Herk, L. Janssens, R. Willems, H. Heidbuchel, and A. E. Aubert, "Determination of atrial fibrillation frequency using QRST-cancellation with QRS-scaling in standard electrocardiogram leads," in *Proc. Comput. Cardiol.*, vol. 32, pp. 339–342, IEEE Press, 2005.

[14] N. V. Thakor and Z. Yi-Sheng, "Applications of adaptive filtering to ECG analysis: Noise cancellation and arrhythmia detection," *IEEE Trans. Biomed. Eng.*, vol. 38, pp. 785–794, 1991. DOI: 10.1109/10.83591

[15] P. Laguna, R. Jané, E. Masgrau, and P. Caminal, "The adaptive linear combiner with a periodic-impulse reference input as a linear comb filter," *Signal Proc.*, vol. 48, pp. 193–203, 1996. DOI: 10.1016/0165-1684(95)00135-2

[16] L. Sörnmo and P. Laguna, *Bioelectrical Signal Processing in Cardiac and Neurological Applications.* Amsterdam: Elsevier (Academic Press), 2005.

[17] J. Malmivuo and R. Plonsey, *Bioelectromagnetism.* Oxford: Oxford University Press, 1995.

[18] G. J. M. Huiskamp and A. van Oosterom, "Heart position and orientation in forward and inverse electrocardiography," *Med. Biol. Eng. & Comput.*, vol. 30, pp. 613–620, 1992. DOI: 10.1007/BF02446793

[19] M. Stridh and L. Sörnmo, "Spatiotemporal QRST cancellation techniques for analysis of atrial fibrillation," *IEEE Trans. Biomed. Eng.*, vol. 48, pp. 105–111, 2001. DOI: 10.1109/10.900266

[20] G. H. Golub and C. F. van Loan, *Matrix Computations.* Baltimore: The Johns Hopkins University Press, 2nd ed., 1989.

[21] J. Waktare, K. Hnatkova, C. J. Meurling, H. Nagayoshi, T. Janota, A. J. Camm, and M. Ma-lik, "Optimal lead configuration in the detection and subtraction of QRS and T wave templates in atrial fibrillation," in *Proc. Comput. Cardiol.*, pp. 629–632, IEEE Press, 1998. DOI: 10.1109/CIC.1998.731952

[22] M. Lemay, J.-M. Vesin, A. van Oosterom, V. Jacquemet, and L. Kappenberger, "Cancellation of ventricular activity in the ECG: Evaluation of novel and existing methods," *IEEE Trans. Biomed. Eng.*, vol. 54, pp. 542–546, 2007. DOI: 10.1109/TBME.2006.888835

[23] A. van Oosterom, "The dominant T wave and its significance," *J. Cardiovasc. Electrophysiol.*, vol. 14, pp. S180–187, 2003. DOI: 10.1046/j.1540.8167.90309.x

[24] A. van Oosterom, "The dominant T wave," *J. Electrocardiol.*, vol. 37, pp. 193–197, 2004. DOI: 10.1016/j.jelectrocard.2004.08.056

[25] W. H. Press, S. A. Teukolsky, W. T. Vetterling, and B. P. Flannery, *Numerical Recipes in C: The Art of Scientific Computing*. New York: Cambridge Univ. Press, 2nd ed., 1992. DOI: 10.2277/0521431085

[26] I. T. Joliffe, *Principal Component Analysis*. Berlin: Springer Verlag, 2002.

[27] F. Castells, P. Laguna, L. Sörnmo, A. Bollmann, and J. Millet, "Principal component analysis in ECG signal processing," *J. Adv. Signal Proc. (www.hindawi.com/journals/asp)*, vol. 2007, no. ID 74580, 2007. DOI: 10.1155/2007/74580

[28] F. Castells, C. Mora, J. J. Rieta, D. Moratal-Pérez, and J. Millet, "Estimation of atrial fibrillatory wave from single-lead atrial fibrillation electrocardiograms using principal component analysis concepts," *Med. Biol. Eng. & Comput.*, vol. 43, pp. 557–560, 2005. DOI: 10.1007/BF02351028

[29] P. Langley, J. P. Bourke, and A. Murray, "Frequency analysis of atrial fibrillation," in *Proc. Comput. Cardiol.*, vol. 27, pp. 65–68, IEEE Press, 2000. DOI: 10.1109/CIC.2000.898456

[30] D. Raine, P. Langley, A. Murray, A. Dunuwille, and J. P. Bourke, "Surface atrial frequency analysis in patients with atrial fibrillation: A tool for evaluating the effects of intervention," *J. Cardiovasc. Electrophysiol.*, vol. 15, pp. 1021–1026, 2004. DOI: 10.1046/j.1540-8167.2004.04032.x

[31] J. F. Cardoso, "Blind signal separation: Statistical principles," *Proc. IEEE*, vol. 86, pp. 2009–2025, 1998. DOI: 10.1109/5.720250

[32] A. Hyvärinen, J. Karhunen, and E. Oja, *Independent Component Analysis*. Wiley Interscience, 2001.

[33] J. J. Rieta, F. Castells, C. Sánchez, V. Zarzoso, and J. Millet, "Atrial activity extraction for atrial fibrillation analysis using blind source separation," *IEEE Trans. Biomed. Eng.*, vol. 51, pp. 1176–1186, 2004. DOI: 10.1109/TBME.2004.827272

[34] M. Lemay, J.-M. Vesin, Z. Ihara, and L. Kappenberger, "Suppression of ventricular activity in the surface electrocardiogram of atrial fibrillation," in *Proc. ICA*, pp. 1095–1102, 2004. DOI: 10.1007/b100528

[35] F. Castells, J. J. Rieta, J. Millet, and V. Zarzoso, "Spatiotemporal blind source separation approach to atrial activity estimation in atrial tachyarrhythmias," *IEEE Trans. Biomed. Eng.*, vol. 52, pp. 258–267, 2005. DOI: 10.1109/TBME.2004.840473

[36] R. Phlypo, Y. D'Asseler, I. Lemahieu, and V. Zarzoso, "Extraction of the atrial activity from the ECG based on independent component analysis with prior knowledge of the source kurtosis signs," in *Proc. IEEE EMBS*, pp. 6499–6502, 2007. DOI: 10.1109/IEMBS.2007.4353848

[37] G. D. Clifford, F. Azuaje, and P. E. McSharry, eds., *Advanced Methods and Tools for ECG Data Analysis*. Boston: Artech House, 2006.

[38] A. J. Pullan, M. L. Buist, and L. K. Cheng, *Mathematically Modelling the Electrical Activity of the Heart*. Hackensack, NJ: World Scientific, 2005.

[39] C. Vayá, J. J. Rieta, C. Sanchez, and D. Moratal, "Convolutive blind source separation algorithms applied to the electrocardiogram of atrial fibrillation: Study of performance," *IEEE Trans. Biomed. Eng.*, vol. 54, pp. 1530–1533, 2007. DOI: 10.1109/TBME.2006.889778

[40] A. Belouchrani, K. Abed-Meraim, J. F. Cardoso, and E. Moulines, "A blind source separation technique using second-order statistics," *IEEE Trans. Signal Proc.*, vol. 45, pp. 434–444, 1997. DOI: 10.1109/78.554307

[41] F. Sandberg, M. Stridh, and L. Sörnmo, "Robust time–frequency analysis of atrial fibrillation using hidden Markov models," *IEEE Trans. Biomed. Eng.*, vol. 55, pp. 502–511, 2008. DOI: 10.1109/TBME.2007.905488

[42] V. D. A. Corino, L. T. Mainardi, M. Stridh, and L. Sörnno, "Improved time–frequency analysis of atrial fibrillation signals using spectral modelling," *IEEE Trans. Biomed. Eng.*, vol. 55, 2008 (in press).

[43] V. Jacquemet, A. van Oosterom, J.-M. Vesin, and L. Kappenberger, "Analysis of electrocardiograms during atrial fibrillation," *IEEE Med. Biol. Eng. Mag.*, vol. 25, pp. 79–88, 2006. DOI: 10.1109/EMB-M.2006.250511

[44] P. Langley, M. Stridh, J. J. Rieta, J. Millet, L. Sörnmo, and A. Murray, "Comparison of atrial signal extraction algorithms in 12-lead ECGs with atrial fibrillation," *IEEE Trans. Biomed. Eng.*, vol. 53, pp. 343–346, 2006. DOI: 10.1109/TBME.2005.862567

CHAPTER 4

Time–Frequency Analysis of Atrial Fibrillation

Frida Sandberg, Martin Stridh, and Leif Sörnmo

4.1 INTRODUCTION

Estimation of the atrial fibrillatory frequency, i.e., the repetition rate of the f-waves, is an important goal when analyzing body surface ECG recordings with atrial fibrillation (AF). Such analysis requires that the atrial activity has first been extracted by some suitable method, resulting in an atrial signal; see Chapter 3. By comparing endocardially recorded electrograms with ECGs, it has been established that the ECG-based AF frequency estimate can be used as an index of the length of the average atrial fibrillatory cycle [1, 2]. Atrial fibrillation with low AF frequency is more likely to terminate spontaneously and to respond better to antiarrhythmic drugs or cardioversion, whereas high AF frequency is more often associated with persistence to therapy. Several studies have demonstrated significant correlation between AF frequency and the likelihood of spontaneous or drug-induced AF termination [3]; in particular, low AF frequency may serve as a predictor of spontaneous AF termination [4]. The likelihood of successful pharmaceutical cardioversion is higher when the AF frequency is below 6 Hz [2, 5]. Moreover, the risk of early AF recurrence is higher for patients with higher AF frequency [6] and, therefore, AF frequency may be taken into consideration when selecting candidates for cardioversion.

Power spectral analysis of the atrial signal is the most common approach to locate AF frequency: the location of the largest spectral peak is taken as the AF frequency estimate. Most clinical studies make use of nonparametric, Fourier-based spectral analysis in which the atrial signal is divided into shorter, overlapping segments, where each segment is subjected to windowing, e.g., using Welch's method [7]. The desired power spectrum is obtained by averaging the power spectra of the respective segments. It is customary to pad each segment with zeros so that the location of the spectral peak can be determined more precisely (although zero padding does not improve spectral resolution per se).

Segment length is likely to be the most important design parameter in nonparametric spectral analysis since it determines the estimation accuracy of AF frequency by restricting spectral resolution. It is advisable that the segment length is chosen to be at least a few seconds so as to produce an acceptable variance of the power spectrum. Concerning spectral resolution, it may seem warranted to use longer segments in order to assure that concurrent, slightly different rates are appropriately resolved. For example, a 10-s segment of the atrial signal yields, at best, a frequency resolution of 0.1 Hz, depending on the chosen window. However, little evidence has been put forward in the literature which supports the existence of concurrent fibrillatory rates, and, therefore, spectral resolution is less crucial to consider when choosing segment length. Figure 4.1 displays the power spectra obtained from a three-lead atrial signal. In all leads, the largest spectral peak occurs at approximately the same location; lead V_1 contains the f-waves with largest amplitude as the spectral peak is largest in this lead.

Since the frequencies which characterize the atrial signal are confined to an interval well below 25–30 Hz, a sampling rate much lower than the one used for extraction of the atrial signal should be used to reduce the amount of computations. Thus, an original sampling rate of the ECG signal of 1,000 Hz may be decimated to as low as 50 Hz without loss of information before spectral analysis. Although sampling rate decimation is dispensable when nonparametric spectral analysis is performed, it is crucial to perform when parametric spectral estimation techniques are employed in order to avoid the presence of spurious spectral peaks [8].

Power spectral analysis reflects the average signal behavior during the analyzed time interval, the location of the dominant spectral peak being the main information carrier with clinical significance. However, it is well known that this analysis suffers from an inability to characterize temporal variations in AF frequency. The presence of bi- or multimodal spectral peaks, reported on in some early studies [1, 9], can just as well be explained by temporal variations in AF frequency as by spatial variations. From an electrophysiological viewpoint, there are good reasons to believe that the atrial fibrillatory waves have time-dependent properties since they reflect complex patterns of electrical activation wavefronts. Therefore, it is advisable to employ time–frequency analysis in order to track variations in AF frequency when more detailed information is needed.

Time–frequency analysis is a powerful tool for unveiling temporal variations of the AF signal, whether such variations are spontaneous or due to intervention [10, 11]. The AF frequency is known to be influenced by autonomic modulation and therefore its variations over time has been studied in terms of the effects of parasympathetic and sympathetic stimulation as well as with respect to circadian rhythm. It has been shown that AF frequency decreases during the night and increases in the morning [12, 13], and that both carotid sinus massage (parasympathetic stimulation) [14] and head-up tilt (sympathetic stimulation) [15] alters AF frequency.

While a plethora of methods has been developed for time–frequency analysis, the method that provides the most adequate characterization of the atrial signal is yet to be determined. The Wigner–Ville distribution (WVD) and the cross Wigner–Ville distribution (XWVD) are two of the first approaches which were considered for time–frequency analysis of AF signals (Section 4.2). More recently, the spectral profile method has been suggested for such analysis, being especially designed to account for the harmonic structure of AF signals (Section 4.3). In situations where the ECG signal is particularly noisy, e.g., when recorded during ambulatory conditions or stress testing,

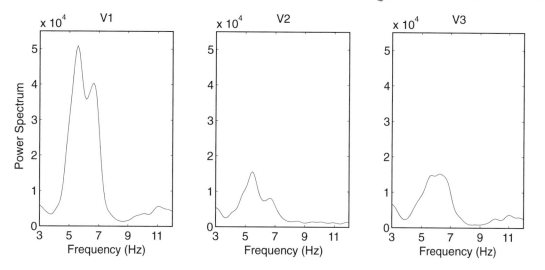

Figure 4.1: Power spectra of a three-lead AF signal (leads V1, V2, and V3), exhibiting peaks at about 6 Hz. In general, the AF power spectrum has a dominant peak within the interval 3–12 Hz, the peak location determining the average fibrillatory rate of nearby endocardial sites. Depending on the shape of the f-waves, the spectrum may exhibit one or several harmonics with magnitude lower than that of the fundamental frequency, i.e., the AF frequency.

it is highly desirable to develop approaches which robustify the time–frequency analysis. Section 4.4 presents various robust approaches where concepts such as spectral validity, spectral modeling using Gaussian functions, and hidden Markov models are crucial.

4.2   TIME–FREQUENCY DISTRIBUTIONS

In its simplest form, time–frequency analysis can be performed by dividing the continuous-time atrial signal $x(t)$ into short, consecutive, possibly overlapping segments, each segment being subjected to spectral analysis. The resulting series of spectra reflects the time-varying nature of the signal. The most common approach to time–frequency analysis is the nonparametric, Fourier-based spectral analysis applied to each segment—an operation which is known as the short-term Fourier transform (STFT). In that approach, the definition of the Fourier transform is modified so that a sliding time window, denoted $w(t)$, defines each time segment to be analyzed. The result is a two-dimensional function $X(t, \Omega)$,

$$X(t, \Omega) = \int_{-\infty}^{\infty} x(\tau)w(\tau - t)e^{-j\Omega\tau}d\tau \,, \tag{4.1}$$

where $\Omega = 2\pi F$ denotes analog frequency and $w(t)$ is positive-valued. The length of $w(t)$ determines the resolution in time and frequency in such a way that a short window yields good time resolution but poor frequency resolution, whereas a long window yields the opposite. Analogous to the computation of the classical periodogram, the spectrogram of $x(t)$ is obtained by computing the squared magnitude of the STFT in (4.1),

$$S_x(t, \Omega) = |X(t, \Omega)|^2 \ . \tag{4.2}$$

Thus, the spectrogram is a real-valued, nonnegative distribution which provides a signal representation in the time–frequency domain.

It may be desirable to track changes in AF frequency as fine as 0.1 Hz and, therefore, as mentioned above, segments with at least 10-s length are needed to achieve that frequency resolution. On the other hand, AF frequency may change quite rapidly such that a time resolution of at least a few seconds is desirable; this is illustrated in Figure 4.2. These conflicting requirements have proven difficult to achieve when using the STFT and, therefore, other techniques for time–frequency analysis have to be investigated which offer improved resolution.

4.2.1   THE WIGNER–VILLE DISTRIBUTION

While the STFT depends linearly on the signal $x(t)$, there are other time–frequency distributions which depend quadratically, thereby offering improved resolution in time and frequency. The WVD is one such distribution, defined by [19, 20]

$$W_x(t, \Omega) = \frac{1}{2\pi} \int_{-\infty}^{\infty} \int_{-\infty}^{\infty} A_x(\tau, \nu) e^{-j\nu t} e^{-j\Omega\tau} d\nu d\tau \ , \tag{4.3}$$

where $A_x(\tau, \nu)$ denotes the ambiguity function, defined by

$$A_x(\tau, \nu) = \int_{-\infty}^{\infty} x^*(t - \frac{\tau}{2}) x(t + \frac{\tau}{2}) e^{j\nu t} dt \ , \tag{4.4}$$

where * denotes the complex conjugate. The ambiguity function is a two-dimensional function which reflects the uncertainty in time and frequency associated with $x(t)$. This function can be understood as the Fourier transform in the variable t of the product $x(t - \frac{\tau}{2}) x^*(t + \frac{\tau}{2})$, describing the deterministic, instantaneous correlation of two values separated by the time lag τ. An alternative expression of the WVD definition is obtained by combining (4.3) and (4.4),

$$W_x(t, \Omega) = \int_{-\infty}^{\infty} x^*(t - \frac{\tau}{2}) x(t + \frac{\tau}{2}) e^{-j\Omega\tau} d\tau \ . \tag{4.5}$$

It should be noted that the analytic signal of $x(t)$ is used instead of the original signal when computing the WVD or other quadratic time–frequency distributions (although the notation $x(t)$ is still being kept) [20].

A problem with quadratic distributions of multi-component signals is the introduction of cross-terms. In the ambiguity domain, defined by $A_x(\tau, \nu)$, the auto-terms are concentrated at the origin, whereas the cross-terms are located further away from the origin. A general class of time–frequency distributions has been introduced whose degrees of freedom can be exploited to mitigate

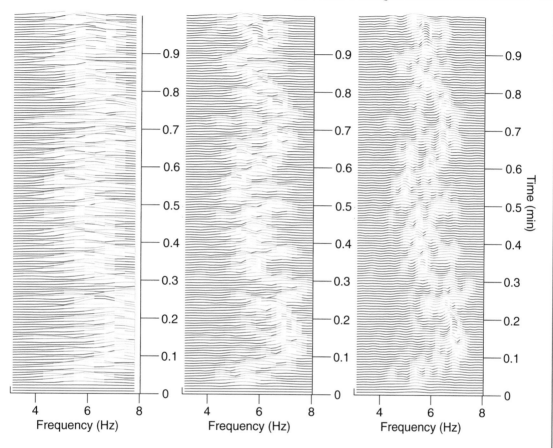

Figure 4.2: The short-term Fourier transform of a 60-s atrial signal computed for the segment lengths 1.28, 2.56, and 5.12 s (left to right), corresponding to 64, 128, and 256 points, respectively, at 50-Hz sampling rate. Visual inspection suggests that 2.5 s length offer a reasonable trade-off between time and frequency resolution; this particular length has been used in many clinical studies, see, e.g., [16, 17, 18].

the cross-term problem [21]. A two-dimensional kernel function $g(\tau, v)$ weighs the ambiguity function such that undesired cross-terms being far away from the origin are suppressed, whereas the auto-terms remain essentially unaffected. The general time–frequency distribution is defined as

$$C_x(t, \Omega) = \frac{1}{2\pi} \int_{-\infty}^{\infty} \int_{-\infty}^{\infty} g(\tau, v) A_x(\tau, v) e^{-Jvt} e^{-J\Omega\tau} dv d\tau \tag{4.6}$$

which is known as Cohen's class [20].

A large number of kernel functions has been presented [22], and one of the most popular members of Cohen's class is the Choi–Williams distribution (CWD). This time–frequency distribution, which has been extensively applied in biomedical signal processing, is defined by the exponential kernel [23, 24]

$$g(\tau, \nu) = e^{-\nu^2 \tau^2 / (4\pi^2 \sigma)}, \quad \sigma > 0,\tag{4.7}$$

where σ is a design parameter that determines the degree of cross-term interference reduction and related smoothing effect. For a small value of σ, the kernel is concentrated around the origin in the ambiguity domain (except for the τ and ν axes), whereas the resulting distribution tends to the WVD for large values of σ. With a suitable choice of σ, the kernel can be used to reduce the influence of cross-terms, while essentially preserving the auto-terms.

It has been suggested that the CWD is suitable for long-term analysis of atrial signals in which the purpose is to track slow changes in AF frequency [10]; here, "slow" signifies changes which take place over minutes, whereas "fast" signifies second-to-second changes. The long-term AF frequency trend can be estimated by computing the "center of gravity" of $W_x(t, \Omega)$ [10], see also [25],

$$\hat{F}(t) = \frac{\displaystyle\int_{\Omega_l}^{\Omega_u} \Omega W_x(t, \Omega) d\Omega}{\displaystyle\int_{\Omega_l}^{\Omega_u} W_x(t, \Omega) d\Omega},\tag{4.8}$$

where Ω_l and Ω_u define the lower and upper limit of the frequency interval, respectively, in which the AF frequency is to be located; typically, this interval ranges from 3–12 Hz. The estimator in (4.8) is more robust than is straightforward peak detection, however, at the expense of the resulting estimate being more biased.

The CWD has been considered in some preliminary studies where the aim was to study and quantify the effect of antiarrhythmic drugs, i.e., sotalol [10] and amiodarone [26]. Another application has been the detection of changes in AF frequency that may accompany changes in autonomic tone, e.g., observed during a head-up tilt test; this maneuver is known to increase sympathetic discharge [10]. The CWD has also been considered for distinguishing atrial flutter from paroxysmal AF with different degrees of organization [27].

4.2.2   THE CROSS WIGNER–VILLE DISTRIBUTION

Another approach to the estimation of the AF frequency trend is to employ the XWVD as it offers the advantage of integrating trend estimation with the computation of the WVD. The XWVD offers good performance for noisy signals which are long relative the window length [28]. The general definition of the XWVD is given by [29]

$$W_{x_1, x_2}(t, \Omega) = \int_{-\infty}^{\infty} x_1^*(t - \frac{\tau}{2}) x_2(t + \frac{\tau}{2}) e^{-j\Omega\tau} d\tau,\tag{4.9}$$

where $x_1(t)$ and $x_2(t)$ are two arbitrary analytical signals. With the goal to estimate the frequency trend in the observed signal $x(t)$, $x_1(t)$ and $x_2(t)$ are replaced by $x(t)$ and a unit amplitude, frequency-modulated sinusoid $x_F(t)$, respectively. The latter signal is defined by

$$x_F(t) = \exp\left[j2\pi \int_{-\infty}^{t} F(\tau)d\tau \right] , \tag{4.10}$$

where $F(t)$ denotes frequency trend.

An iterative procedure is employed to find $F(t)$ such that successive trend estimates are obtained from the peak of the XWVD. The initial trend estimate, denoted $\hat{F}^{(0)}(t)$, may be determined from, e.g., the largest peak of the STFT. The procedure includes the following three steps, starting with the iteration index l set to zero.

1. The unit amplitude, frequency-modulated sinusoid is computed from the frequency trend $\hat{F}^{(l)}(t)$

$$\hat{x}_F^{(l)}(t) = \exp\left[j2\pi \int_{-\infty}^{t} \hat{F}^{(l)}(\tau)d\tau \right] . \tag{4.11}$$

2. The XWVD between $x(t)$ and $x_F(t)$ is computed,

$$W_{x,x_F}^{(l)}(t, \Omega) = \int_{-\infty}^{\infty} x^*(t - \frac{\tau}{2})\hat{x}_F^{(l)}(t + \frac{\tau}{2})e^{-j\Omega t}d\tau , \tag{4.12}$$

and the peak location is extracted from $W_{x,x_F}^{(l)}(t, \Omega)$ so that a new frequency trend estimate $\hat{F}^{(l+1)}(t)$ results.

3. The procedure is repeated from step 1 until the difference between $\hat{F}^{(l+1)}(t)$ and $\hat{F}^{(l)}(t)$, measured by some suitable metric, drops below a certain tolerance value.

It can be shown that if the true signal is identical to a linear FM signal and no noise is present, convergence will occur for an arbitrary initial estimate of $F(t)$ [29]. When analyzing AF signals, convergence has been reported to occur after a few iterations [10]. However, convergence can be very slow, and sometimes not even achieved, in signal segments with low signal-to-noise ratio (SNR) and, therefore, the maximum number of iterations is usually limited. Such signal segments should be excluded from further analysis.

Using the XWVD-based method, considerable temporal variation of the AF frequency is found in a majority of signals, thus explaining the broad or bimodal spectral peaks which sometimes are observed in the power spectrum of a 60-s recording, cf. the example in Figure 4.1. The value of using high temporal resolution is further illustrated by the following two situations: in certain patients the typical, irregular rhythm of AF was intermittently interrupted by short intervals of very regular, flutter-like rhythm. Such intervals were found in the frequency trends produced by the XWVD-based method; see Figure 4.3 [10].

4.3 THE SPECTRAL PROFILE METHOD

Of the different approaches to time–frequency analysis, the so-called spectral profile method is the one which has received the most attention in clinical studies. This method grew out of experience

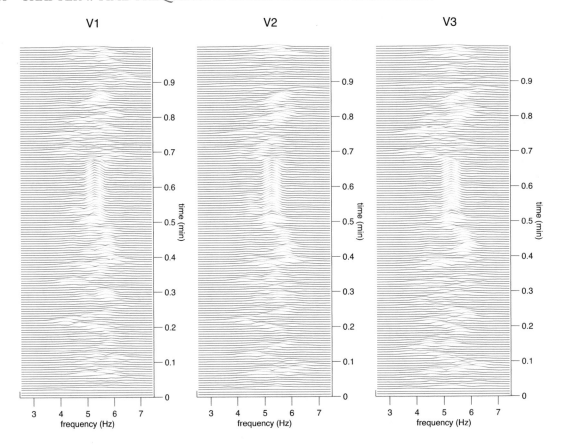

Figure 4.3: Short-term analysis of a 60-s AF signal where the atrial fibrillation has larger amplitude and more stable rate during a 10-s interval starting at about 30 s. The time–frequency distribution was obtained using the cross Wigner–Ville distribution.

from the XWVD approach and the limitation of only considering the fundamental frequency but not the harmonics of the atrial signal. Inclusion of harmonics may not only improve the estimation of the fundamental frequency, but the harmonic pattern may in itself convey information of clinical significance. Using nonparametric, Fourier-based spectral analysis of electrogram recordings, the significance of harmonics has been established by the introduction of a so-called organization index. This index, defined as the ratio between the area under the harmonics and the total area, can be used as a measure that reflects the degree of atrial organization [30, 31]. In particular, the organization index was found useful to predict the outcome of atrial defibrillation: a successful shock showed a dominant fundamental frequency with a distinct harmonic pattern, whereas an unsuccessful shock showed a dominant fundamental frequency but without any clear harmonic pattern.

Using the spectral profile method, a time–frequency distribution results from the local power spectra of successive short segments (similar to the STFT). Next, the distribution is decomposed into a spectral profile and a number of trends describing variations in AF frequency as well as in f-wave morphology [11]. The spectral profile resembles the conventional power spectrum, however, frequency shifting and amplitude scaling are performed before spectral averaging. As a result, the peaks of the spectral profile become more distinct, implying that determination of the harmonic structure of the atrial signal, reflecting f-wave morphology, is facilitated.

In mathematical terms, the spectral profile method divides the atrial signal into overlapping segments $\mathbf{x}_l$ for which the corresponding spectrum $\mathbf{q}_l$ is obtained with a nonuniform Fourier transform,

$$\mathbf{q}_l = \mathbf{F}\mathbf{W}\mathbf{x}_l \,, \tag{4.13}$$

where the elements of the $N \times N$ diagonal matrix $\mathbf{W}$ define a window function. The vectors $\mathbf{x}_l$ and $\mathbf{q}_l$ are $N \times 1$ and $K \times 1$, respectively. The $K \times N$ matrix $\mathbf{F}$ defines the K-point discrete, nonuniform Fourier transform,

$$\mathbf{F} = \begin{bmatrix} \mathbf{1} & e^{-j2\pi\mathbf{f}} & e^{-j2\pi\mathbf{f}2} & \cdots & e^{-j2\pi\mathbf{f}(N-1)} \end{bmatrix} \,, \tag{4.14}$$

where $\mathbf{f} = \begin{bmatrix} f_0 & \cdots & f_{K-1} \end{bmatrix}^T$ is a logarithmically defined frequency vector, i.e.,

$$f_k = f_0 \cdot 10^{\frac{k}{K}}, \quad k = 0, \ldots, K - 1 \,, \tag{4.15}$$

and $\mathbf{1}$ a column vector of length K with all elements equal to one. It should be noted that f denotes digital frequency, whereas F denotes analog frequency, see, for example, (4.1). In this method, the phase of $\mathbf{q}_l$ is ignored and therefore $\mathbf{q}_l$ is below redefined to

$$|\mathbf{q}_l| \to \mathbf{q}_l \,. \tag{4.16}$$

The logarithmic frequency scale is employed because doubling in frequency for the entire spectrum corresponds to the same number of frequency bins and, therefore, the harmonic pattern of two spectra with different fundamental frequency can be matched. This property is important when estimating the frequency shift of each $\mathbf{q}_l$ that best matches the spectral profile.

Each spectrum $\mathbf{q}_l$ is assumed to be modeled by $\mathbf{s}_l$ which is a frequency-shifted and amplitude-scaled version of a known spectral profile $\boldsymbol{\phi}_l$,

$$\mathbf{s}_l = a_l \mathbf{J}_{\theta_l} \boldsymbol{\phi}_l \,, \tag{4.17}$$

where the matrix $\mathbf{J}_{\theta_l}$ handles frequency shifting by selecting an appropriate interval of points from the vector $\boldsymbol{\phi}_l$. The frequency-shift parameter θ_l and the amplitude scaling parameter a_l are estimated by minimizing the quadratic cost function $J(\theta_l, a_l)$ so that the model spectrum $\mathbf{s}_l$ matches $\mathbf{q}_l$ optimally,

$$J(\theta_l, a_l) = (\mathbf{q}_l - \mathbf{s}_l)^T \mathbf{D} (\mathbf{q}_l - \mathbf{s}_l) \,. \tag{4.18}$$

The diagonal weight matrix $\mathbf{D}$ is designed to weight the error of the frequency components differently so as to compensate for the logarithmic frequency scale; details on the design of $\mathbf{D}$ can be found in [11]. Joint minimization of (4.18) with respect to θ_l and a_l results in the following two estimators:

$$\hat{\theta}_l = \arg \max_{\theta_l} \left[\mathbf{q}_l^T \mathbf{D}^{\frac{1}{2}} \mathbf{J}_{\theta_l} \mathbf{D}^{\frac{1}{2}} \boldsymbol{\phi}_l \right] , \tag{4.19}$$

$$\hat{a}_l = \mathbf{q}_l^T \mathbf{D}^{\frac{1}{2}} \mathbf{J}_{\hat{\theta}_l} \mathbf{D}^{\frac{1}{2}} \boldsymbol{\phi}_l . \tag{4.20}$$

Since the spectral profile is not known a priori, a self-tuning method has been employed to estimate the spectral profile from the prevailing signal characteristics. Exponential averaging has been found useful for this purpose, defined by the recursion

$$\hat{\boldsymbol{\phi}}_{l+1} = (1 - \alpha_l)\hat{\boldsymbol{\phi}}_l + \alpha_l \frac{\mathbf{J}_{-\hat{\theta}_l}\tilde{\mathbf{q}}_l}{\|\mathbf{J}_{-\hat{\theta}_l}\tilde{\mathbf{q}}_l\|} , \quad l \geq 0 , \tag{4.21}$$

where the gain α_l is set to a positive value ($0 < \alpha_l < 1$) when the signal is judged to be reliable, otherwise it is set to zero. Here, the notation $(\tilde{\cdot})$ indicates that $\mathbf{q}_l$ has been pre- and appended with a sufficient number of points in order to allow for frequency shifting. The initial spectral profile $\hat{\boldsymbol{\phi}}_0$ can be defined to have values close to zero except for one element where it is set to one; this element is positioned in an interval where the AF frequency is likely to be found.

A point in the AF frequency trend is obtained by determining the peak location in the spectral profile $\hat{\boldsymbol{\phi}}_l$ and then correcting it with $\hat{\theta}_l$. The amplitude estimate $\hat{a}_l$ can be used as a measure that reflects f-wave amplitude. It can also serve as a normalization factor when evaluating the model error $J(\hat{\theta}_l, \hat{a}_l)$.

The f-wave morphology can be characterized by the decay with which the magnitudes of the harmonics drop off relative to the magnitude of the fundamental frequency. The exponential decay can be estimated by fitting a spectral line model to $\hat{\boldsymbol{\phi}}_l$, only considering harmonics which are below 20 Hz [11]. A closed-form expression of the exponential decay estimator can be obtained when a least squares approach is adopted. With the spectral line model, a wide range of shapes can be represented, spanning from the characteristic "sawtooth-like shape" of organized atrial flutter/fibrillation and tachycardia (slow exponential decay) to less organized and more sinusoidal-looking AF (fast exponential decay).

Two different examples of AF frequency estimation using the spectral profile method are presented in Figure 4.4. From the example in Figure 4.4(a), it is evident that the spectral profile offers better characterization of the harmonic structure of the atrial signal than does the conventional magnitude spectrum. This property is due to the frequency shifting operation combined with spectral averaging.

4.4   ROBUST SPECTRAL ESTIMATION

In situations when the f-wave amplitude is weak or the atrial signal is corrupted with noise, time–frequency analysis using the spectral profile method becomes less reliable. The noise may originate

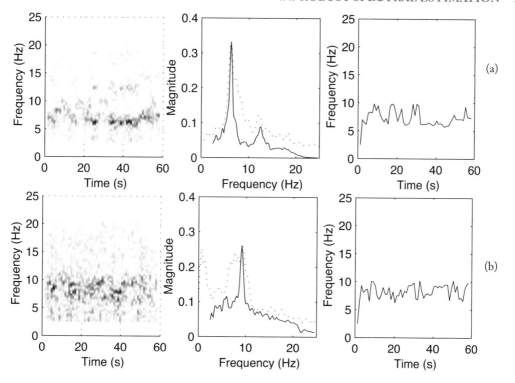

Figure 4.4: Time–frequency analysis of two 60-s atrial signals using the spectral profile method, resulting in a logarithmic distribution (left panel), the spectral profile and the conventional magnitude spectrum (solid and dotted lines, respectively) (middle panel), and the fibrillation frequency trend (right panel). (a) A spectral profile with a relatively large first harmonic, and (b) a spectral profile where the harmonics are lacking. In both cases, the variations in AF frequency are quite large.

from extracardiac sources as well as from poor QRST cancellation. Various approaches have been suggested to handle such situations, for example, by testing the validity of the spectral profile before it is subjected to feature extraction (Section 4.4.1). In another approach, each local spectrum $\mathbf{q}_l$ can be tested before being included in the update of the spectral profile. The test is based on a spectral model of the atrial signal which is fitted to each local spectrum; the resulting model error then serves as a measure of reliability (Section 4.4.2). Yet another approach is to exclude, or possibly change, frequency estimates that do not fit the AF frequency trend, assuming that variations in AF frequency can be characterized by a hidden Markov model (Section 4.4.3).

4.4.1   SPECTRAL PROFILE VALIDITY
The presence of AF implies that the spectral profile should exhibit a certain harmonic structure in order to be considered for further analysis, e.g., the computation of the above-mentioned exponential

decay. An harmonic structure may be ensured by implementing a number of tests which are described in the following [32].

The spectral profile is represented by the vector

$$
\boldsymbol{\phi}_l = \begin{bmatrix} \phi_l(1) \\ \phi_l(2) \\ \vdots \\ \phi_l(K) \end{bmatrix},
\tag{4.22}
$$

where $\phi_l(k)$ denotes the kth frequency bin, l signal segment index, and K the total number of bins in $\boldsymbol{\phi}_l$. In order to judge the validity of $\boldsymbol{\phi}_l$, four parameters are considered, namely, the spectral SNR $z_{SNR}(l)$, the second peak position $z_{SPP}(l)$, the second peak amplitude $z_{SPM}(l)$, and the model error $J(\hat{\theta}_l, \hat{a}_l)$, cf. (4.18).

The spectral SNR is defined as the ratio between the mean magnitude of the fundamental frequency and the first harmonic, denoted $\phi_l(p_0(l))$ and $\phi_l(p_1(l))$, respectively, and the magnitude of the background noise $\phi_l(p_b(l))$, i.e.,

$$
z_{SNR}(l) = \frac{\phi_l(p_0(l)) + \phi_l(p_1(l))}{2\phi_l(p_b(l))}.
\tag{4.23}
$$

The positions of the fundamental frequency and the first harmonic are denoted $p_0(l)$ and $p_1(l)$, respectively. The background noise position $p_b(l)$ is taken to be halfway in between the fundamental frequency and the first harmonic.

The second peak position is defined as the ratio between the distance from $p_0(l)$ to the second largest peak, positioned at $p_s(l)$, and the distance from $p_0(l)$ to $p_1(l)$, i.e.,

$$
z_{SPP}(l) = \left| \frac{p_0(l) - p_s(l)}{p_0(l) - p_1(l)} \right|.
\tag{4.24}
$$

The motivation for introducing this parameter is that signals without atrial activity above 2.5 Hz, e.g., P waves at a slower rate, often produce a ringing spectral profile. The second largest peak of such a ringing profile is, however, often very close to the largest peak and does not occur at the expected position of the first harmonic. Since the frequency scale is logarithmic, the number of frequency bins between the fundamental and the first harmonic is fixed. The parameter $z_{SPP}(l)$ detects if the second largest peak deviates from the position of the first harmonic.

The second peak magnitude, being defined as the ratio between the magnitude of the second largest peak $\phi_l(p_s(l))$ and the magnitude of the fundamental frequency $\phi_l(p_0(l))$, i.e.,

$$
z_{SPM}(l) = \frac{\phi_l(p_s(l))}{\phi_l(p_0(l))},
\tag{4.25}
$$

detects when the first harmonic is too large. A ringing spectral profile, related to a rhythm whose fundamental frequency is below 2.5 Hz, is associated with too large a value of $z_{SPM}(l)$.

The spectral profile is judged to be valid if [33]:

1. the spectral SNR $z_{SNR}(l)$ exceeds a certain threshold;

2. neither the second largest peak $z_{SPP}(l)$ nor the second peak magnitude $z_{SPM}(l)$ indicates a ringing profile; and

3. the model error $J(\hat{\theta}_l, \hat{a}_l)$ exhibits no sudden decrease from segment to segment.

Figure 4.5(a) displays a spectral profile with valid harmonic structure, whereas the one displayed in Figure 4.5(b) is invalid. In the latter case, the ringing of the spectral profile is reflected by the second largest peak being too large in comparison with the fundamental, i.e., a too large value of $z_{SPP}(l)$. In addition, the second largest peak is not located at the expected position of the first harmonic, i.e., a too small value of $z_{SPP}(n)$. Both $z_{SPP}(l)$ and $z_{SPM}(l)$ were necessary to identify the invalid spectral profile in Figure 4.5(b).

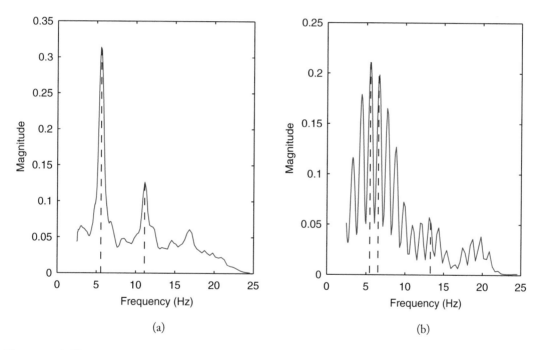

Figure 4.5: Using the validity criteria in Section 4.4.1, the two spectral profiles are correctly judged as being either (a) valid or (b) invalid.

4.4.2   SPECTRAL MODELING WITH GAUSSIAN FUNCTIONS

A disadvantage with the spectral profile method is its lack of control of what goes into the spectral profile: a spectrum reflecting large QRS residuals is just as influential as a spectrum reflecting atrial activity. Although the spectral profile has slow adaptation rate, making it less sensitive to occasional noisy segments, a short sequence of bad segments causes the spectral profile to lose its structure and thereby the AF frequency estimates become invalid. Once the spectral profile has lost its structure, the recovery time until the frequency estimates is valid again can become unacceptably long, even if subsequent segments have an harmonic structure.

These limitations may be remedied by adopting a spectral modeling approach in which the spectrum of each segment is checked before it enters the update of the spectral profile [34]. The model accounts for the fact that the AF spectrum is characterized by a dominant peak at the fundamental frequency f_0 and a number of harmonics centered at approximate integer multiples of f_0. In that study, the AF spectrum was modeled as a sum of Gaussian functions,

$$\varphi(f, \boldsymbol{\theta}) = \sum_{m=0}^{M} A_m e^{-\frac{(f-\mu_m)^2}{2\sigma_m^2}}, \tag{4.26}$$

where $M + 1$ is the number of Gaussians, A_m is the magnitude of the mth Gaussian, σ_m its width, μ_m its location, and $\boldsymbol{\theta}$ a vector containing the model parameters. Since the harmonics are expected to be located at integer multiples of f_0, the location of the mth Gaussian is constrained to an interval centered around $m \cdot f_0$ whose width is defined by the frequency shift Δ_m,

$$\mu_o = f_0, \tag{4.27}$$
$$\mu_m = mf_0 + \Delta_m, \quad m = 1, \ldots, M. \tag{4.28}$$

The parameter vector $\boldsymbol{\theta}$ is defined by

$$\boldsymbol{\theta} = \begin{bmatrix} \mathbf{A} & \boldsymbol{\sigma} & \boldsymbol{\Delta} & f_0 \end{bmatrix}^T, \tag{4.29}$$

where $\mathbf{A} = \begin{bmatrix} A_0 & \cdots & A_M \end{bmatrix}$, $\boldsymbol{\sigma} = \begin{bmatrix} \sigma_0 & \cdots & \sigma_M \end{bmatrix}$, and $\boldsymbol{\Delta} = \begin{bmatrix} \Delta_1 & \cdots & \Delta_M \end{bmatrix}$.

The estimation of model parameters is accomplished by an optimization procedure based on the weighted least squares technique, where the parameter estimates are obtained by minimizing the following cost function:

$$J(\boldsymbol{\theta}_l) = (\mathbf{q}_l - \varphi(\boldsymbol{\theta}_l))^T \mathbf{D}_l (\mathbf{q}_l - \varphi(\boldsymbol{\theta}_l)). \tag{4.30}$$

The vector $\varphi(\boldsymbol{\theta})$ results from evaluating the expression in (4.26) at the logarithmic frequencies f_k,

$$\varphi(\boldsymbol{\theta}) = \begin{bmatrix} \varphi(f_0, \boldsymbol{\theta}) & \cdots & \varphi(f_{K-1}, \boldsymbol{\theta}) \end{bmatrix}. \tag{4.31}$$

Details on the optimization procedure can be found in [34].

A set of parameters which characterizes the harmonic pattern are defined and used to decide whether $\mathbf{q}_l$ should be included in the spectral profile or not. The parameters are descriptors of the spectrum and designed so as to verify if a spectrum exhibits the typical harmonic pattern of AF, i.e., a fundamental component and, possibly, a few harmonics. The exclusion criteria involved the following parameters [34]:

1. the ratio $q(f_m)/q(f_0)$, where $q(f_m)$ identifies the maximum between the fundamental and the first harmonic and $q(f_0)$ the magnitude of the fundamental peak. This ratio reflects the occurrence of spurious peaks between the fundamental and first harmonic;

2. the width σ_0 of the fundamental peak, measuring the spectral concentration of the fundamental; and

3. the model error $J_l(\hat{\boldsymbol{\theta}})$ evaluated only for the peak regions, measuring the similarity between $\mathbf{q}_l$ and $\boldsymbol{\varphi}(\boldsymbol{\theta})$.

Figures 4.6(a) and (b) illustrate the difference in spectral profile with and without application of the exclusion criteria. From these figures, it is obvious that the fundamental peak becomes much clearer when noisy segments have been first excluded. For signals without QRS residuals, the spectral profiles remain unchanged after application of the exclusion criteria as illustrated in Figures 4.6(c) and (d). Another advantage of applying the exclusion criterion is the following. No spectra were excluded from the old spectral profile update, however, a short sequence of noisy segments made the spectral profile unreliable for several subsequent segments. This result was due to the fact that the spectral profile was slowly updated and, consequently, the inclusion of bad spectra had repercussions on the spectral profile in subsequent segments. On the contrary, only a few segments were lost when a bad spectrum was excluded from update with the new method. Thus, for signals with modest presence of QRS residuals, few spectra were excluded and therefore the final spectral profile did not change much (Figures 4.6(c) and (d)). On the other hand, for noisy signals, the number of unreliable segments decreased.

4.4.3 HIDDEN MARKOV MODEL

A hidden Markov model (HMM) may be used as a postprocessing step to make frequency tracking of AF in noisy signals more robust [35]. With this approach, estimates that differ significantly from the frequency trend can be detected and excluded or replaced by estimates based on adjacent frequencies. The Markov model consists of a finite number of states with predefined state transition probabilities. The likelihood of a certain state, corresponding to a unique set of observed variables, depends only on the previous state (random walk). In an HMM, the state variables cannot be directly observed, but each state is associated with certain observation probabilities, i.e., the probabilities of observing a specific set of variables. Given the observation sequence of an HMM, the optimal state sequence can be obtained using the Viterbi algorithm or the forward–backward algorithm [37].

An HMM is well-suited for frequency tracking of the atrial signal. The states of the model correspond to the underlying frequencies, while the observations are determined by the estimated frequency of a signal segment. The HMM for frequency tracking includes a zero state, $z(t) = 0$, when no signal is present, and $P - 1$ different frequency states, $z(t) = 1, \ldots, P - 1$, where state i includes frequencies between f_i and $f_{i+1} = f_i + \Delta f$, with center frequency equal to $\tilde{f}_i$,

$$\tilde{f}_i = f_i + \frac{\Delta f}{2} . \tag{4.32}$$

The HMM is completely characterized by its state transition matrix $\mathbf{A}$, its observation matrix $\mathbf{B}$, and its initial state probability vector $\boldsymbol{\pi}$. The state transition matrix $\mathbf{A}$ describes a priori knowledge of transition probabilities between different states, i.e., different frequency intervals. Its structure

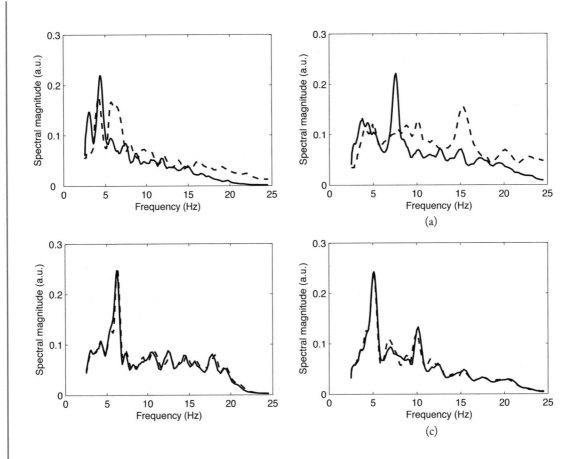

Figure 4.6: (a), (b) Spectral profiles before (dashed line) and after (solid line) application of exclusion criteria. These profiles were obtained from signals with a considerable amount of QRS residuals. (c), (d) Spectral profiles before (dashed line) and after (solid line) application of exclusion criteria, obtained from good quality atrial signals.

depends on the statistical assumptions made on AF frequency. In [35], it was assumed that changes in AF frequency are characterized by a Gaussian probability density function, i.e., the AF frequency is more probable to remain the same or change gradually than to change drastically.

The observation matrix $\mathbf{B}$ describes the probabilities of observing a specific frequency given the true AF frequency, a simple statistical model of the observed atrial signal being the starting point. The matrix elements b_{ij} correspond to the probability of detection in state j when the true state is i. It may be assumed that the observed atrial signal, denoted $s(n)$, is modeled as a sum of harmonically related sinusoids and noise,

$$s(n) = \sum_{k=1}^{K} a_k \sin(2\pi k f_0 n) + v(n) \,. \tag{4.33}$$

The fundamental frequency f_0 and the sinusoidal amplitudes $a_1, a_2, \ldots, a_K$ are assumed to be constant during the signal segment in which the (short-term) Fourier transform is computed. The additive noise $v(n)$ is assumed to be white and Gaussian. The interested reader is referred to [35] for mathematical details on the derivation of the two matrices $\mathbf{A}$ and $\mathbf{B}$.

The Viterbi algorithm finds the most probable underlying frequency trend, given the a priori information of the likelihood of frequency changes in $\mathbf{A}$ and the likelihood of estimating specific frequencies in $\mathbf{B}$ [36, 37]. As a result, frequency estimates that differs from the frequency trend may be excluded or replaced by the HMM, resulting in more robust frequency trend estimation.

The performance of HMM-based frequency tracking was evaluated using the sawtooth simulation model, earlier defined in (3.38). Using this model, the true AF frequency is obviously known. Different types of noise were added to the sawtooth signal, including QRST residuals and muscular artifacts. The QRST residual noise was obtained from ECG signals with normal sinus rhythm, where the QRST complexes were cancelled using average beat subtraction (Section 3.2). The muscular artifact noise was obtained from the MIT-BIH stress test database [38]. The SNR of the simulated signal was defined by

$$\text{SNR} = 20 \log \frac{V_x}{\sigma_v} \,, \tag{4.34}$$

where V_x is the peak-to-peak amplitude of the sawtooth signal, and σ_v is the standard deviation of the noise. Figure 4.7 displays a simulated atrial signal mixed with QRST residual noise.

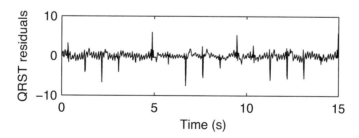

Figure 4.7: Simulated atrial signal mixed with QRST residual noise at an SNR of 5 dB.

Figure 4.8 illustrates frequency tracking in the presence of QRST residual noise when the fundamental frequency of the sawtooth signal is gradually decreasing. At 10 dB SNR, no inaccurate frequency estimates occur and, consequently, the HMM has no effect (Figure 4.8(a)). At 5 dB SNR, two inaccurate frequency estimates are replaced by estimates determined by the HMM, and two inaccurate frequency estimates are excluded by the HMM, i.e., set to the zero state (Figure 4.8(b)).

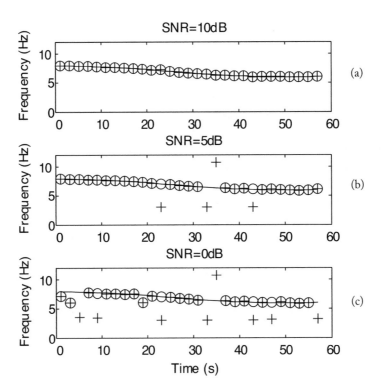

Figure 4.8: Frequency tracking with and without use of the HMM. The simulated signal is obtained for (a) 10 dB SNR, (b) 5 dB SNR, and (c) 0 dB SNR. The true frequency trend (solid line), the estimated frequency using the STFT without HMM ('+'), and the estimated frequency using the STFT with HMM ('o'). Note that absence of '+' or 'o' corresponds to the zero state.

At 0 dB SNR, several inaccurate frequency estimates are either excluded or replaced by frequencies better fitting the trend (Figure 4.8(c)).

4.5   CONCLUSIONS

Power spectral analysis of AF has today found its way into various clinical applications because the AF frequency can be used, e.g., to identify candidates for pharmacological cardioversion and to predict AF termination. While such analysis suffers from inability to characterize the temporal dynamics in AF frequency, the recent development of methods particularly tailored for time–frequency analysis of atrial signals should be helpful in establishing the clinical significance of such temporal information. In this development, special attention has been paid to the fact that the atrial signal during AF often exhibits a poor signal-to-noise ratio and, consequently, requires robust approaches to time–frequency analysis.

Bibliography

[1] M. Holm, S. Pehrsson, M. Ingemansson, L. Sörnmo, R. Johansson, L. Sandhall, M. Sunemark, B. Smideberg, C. Olsson, and S. B. Olsson, "Non-invasive assessment of atrial refractoriness during atrial fibrillation in man—Introducing, validating, and illustrating a new ECG method," *Cardiovasc. Res.*, vol. 38, pp. 69–81, 1998. DOI: 10.1016/S0008-6363(97)00289-7

[2] A. Bollmann, N. Kanuru, K. McTeague, P. Walter, D. B. DeLurgio, and J. Langberg, "Frequency analysis of human atrial fibrillation using the surface electrocardiogram and its response to ibutilide," *Am. J. Cardiol.*, vol. 81, pp. 1439–1445, 1998. DOI: 10.1016/S0002-9149(98)00210-0

[3] A. Bollmann, D. Husser, L. Mainardi, F. Lombardi, P. Langley, A. Murray, J. J. Rieta, J. Millet, S. B. Olsson, M. Stridh, and L. Sörnmo, "Analysis of surface electrocardiograms in atrial fibrillation: Techniques, research, and clinical applications," *Europace*, vol. 8, pp. 911–926, 2006. DOI: 10.1093/europace/eul113

[4] F. Nilsson, M. Stridh, A. Bollmann, and L. Sörnmo, "Predicting spontaneous termination of atrial fibrillation using the surface ECG," *Med. Eng. & Physics*, vol. 26, pp. 802–808, 2006. DOI: 10.1016/j.medengphy.2005.11.010

[5] A. Bollmann, K. Binias, I. Toepffer, J. Molling, C. Geller, and H. Klein, "Importance of left atrial diameter and atrial fibrillatory frequency for conversion of persistent atrial fibrillation with oral flecainide," *Am. J. Cardiol.*, vol. 90, pp. 1011–1014, 2002. DOI: 10.1016/S0002-9149(02)02690-5

[6] J. J. Langberg, J. C. Burnette, and K. K. McTeague, "Spectral analysis of the electrocardiogram predicts recurrence of atrial fibrillation after cardioversion," *J. Electrocardiol.*, vol. 31, pp. 80–84, 1998. DOI: 10.1016/S0022-0736(98)90297-7

[7] M. Hayes, *Statistical Digital Signal Proccessing and Modeling.* New York: John Wiley & Sons, 1996.

[8] R. Sassi, L. T. Mainardi, P. Maison-Blanche, and S. Cerruti, "Estimation of spectral parameters of residual ECG signals during atrial fibrillation using autoregressive models," *Folia Cardiologica*, vol. 12, pp. 108–110, 2005.

[9] S. Pehrson, M. Holm, C. Meurling, M. Ingemansson, B. Smideberg, L. Sörnmo, and S. Olsson, "Non-invasive assessment of magnitude and dispersion of atrial cycle length during chronic atrial fibrillation in man," *Eur. Heart J.*, vol. 19, pp. 1836–1844, 1998. DOI: 10.1053/euhj.1998.1200

[10] M. Stridh, L. Sörnmo, C. J. Meurling, and S. B. Olsson, "Characterization of atrial fibrillation using the surface ECG: Time-dependent spectral properties," *IEEE Trans. Biomed. Eng.*, vol. 48, pp. 19–27, 2001. DOI: 10.1109/10.900245

[11] M. Stridh, L. Sörnmo, C. J. Meurling, and S. B. Olsson, "Sequential characterization of atrial tachyarrhythmias based on ECG time-frequency analysis," *IEEE Trans. Biomed. Eng.*, vol. 51, pp. 100–114, 2004. DOI: 10.1109/TBME.2003.820331

[12] A. Bollmann, K. Sonne, H. Esperer, I. Toepffer, and H. Klein, "Circadian variations in atrial fibrillatory frequency in persistent human atrial fibrillation," *Pacing Clin. Electrophysiol.*, vol. 23, pp. 1867–1871, 2000.

[13] C. J. Meurling, J. E. Waktare, F. Holmqvist, A. Hedman, A. J. Camm, S. B. Olsson, and M. Malik, "Diurnal variations of the dominant cycle length of chronic atrial fibrillation," *Am. J. Physiol.*, vol. 280, pp. H401–H406, 2001.

[14] A. Bollmann, K. Wodarz, H. D. Esperer, I. Toepffer, and H. Klein, "Response of atrial fibrillatory activity to carotide sinus massage in patients with atrial fibrillation," *Pacing Clin. Electrophysiol.*, vol. 24, pp. 1363–1368, 2001. DOI: 10.1046/j.1460-9592.2001.01363.x

[15] M. Ingemansson, M. Holm, and S. B. Olsson, "Autonomic modulation of the atrial cycle length by the head up tilt test: non-invasive evaluation in patients with chronic atrial fibrillation," *Heart*, vol. 80, pp. 71–76, 1998.

[16] D. Husser, M. Stridh, L. Sörnmo, C. Geller, H. U. Klein, S. B. Olsson, and A. Bollmann, "Time–frequency analysis of the surface electrocardiogram for monitoring antiarrhythmic drug effects in atrial fibrillation," *Am. J. Cardiol.*, vol. 95, pp. 526–528, 2005. DOI: 10.1016/j.amjcard.2004.10.025

[17] D. Husser, M. Stridh, D. S. Cannom, A. K. Bhandari, M. J. Girsky, S. Kang, L. Sörnmo, S. B. Olsson, and A. Bollmann, "Validation and clinical application of time–frequency analysis of atrial fibrillation electrocardiograms," *J. Cardiovasc. Electrophysiol.*, vol. 18, pp. 41–46, 2007. DOI: 10.1111/j.1540-8167.2006.00683.x

[18] D. Husser, D. S. Cannom, A. K. Bhandari, M. Stridh, L. Sörnmo, S. B. Olsson, and A. Bollmann, "Electrocardiographic characteristics of fibrillatory waves in new-onset atrial fibrillation," *Europace*, vol. 9, pp. 638–642, 2007. DOI: 10.1093/europace/eum074

[19] S. Qian and D. Chen, *Joint Time–Frequency Analysis. Methods and Applications.* Englewood Cliffs, NJ: Prentice-Hall, 1996. DOI: 10.1109/79.752051

[20] L. Cohen, *Time–Frequency Analysis.* Englewood Cliffs, NJ: Prentice-Hall, 1995.

[21] L. Cohen, "Generalized phase-space distribution functions," *J. Math. Phys.*, vol. 7, pp. 781–786, 1966. DOI: 10.1063/1.1931206

[22] F. Hlawatsch and G. F. Boudreaux-Bartels, "Linear and quadratic time–frequency signal representations," *IEEE Signal Proc. Mag.*, vol. 9, pp. 21–67, 1992. DOI: 10.1109/79.127284

[23] H. Choi and W. J. Williams, "Improved time–frequency representation of multicomponent signals using exponential kernels," *IEEE Acoust. Speech Signal Proc.*, vol. 37, pp. 862–871, 1989. DOI: 10.1109/ASSP.1989.28057

[24] W. J. Williams, "Reduced interference distributions: Biological applications and interpretations," *Proc. IEEE*, vol. 84, pp. 1264–1280, 1996. DOI: 10.1109/5.535245

[25] S. Karlsson, J. Yu, and M. Akay, "Time–frequency analysis of myoelectric signals during dynamic contractions: A comparative study," *IEEE Trans. Biomed. Eng.*, vol. 47, pp. 228–238, 2000. DOI: 10.1109/10.821766

[26] C. Mora, D. Husser, O. Husser, F. Castells, J. Millet, and A. Bollmann, "Systematic evaluation of time–frequency parameters from surface electrocardiograms for monitoring amiodarone effects in atrial fibrillation," in *Proc. Comput. Cardiol.*, vol. 32, pp. 949–952, IEEE Press, 2005.

[27] C. Mora, J. Millet, F. Castells, and R. Ruiz, "Characterization of atrial tachyarrhythmias by means of time–frequency analysis," in *Proc. Comput. Cardiol.*, vol. 31, pp. 437–440, IEEE Press, 2004.

[28] B. Boashash, "Estimating and interpreting the instantaneous frequency of a signal—part 2: Algorithms and applications," *Proc. IEEE*, vol. 80, pp. 540–568, 1992. DOI: 10.1109/5.135378

[29] B. Boashash and P. O'Shea, "Use of the cross Wigner–Ville distribution for estimation of instantaneous frequency," *IEEE Trans. Signal Proc.*, vol. 41, pp. 1439–1445, 1993. DOI: 10.1109/78.205752

[30] T. Everett, L. Kok, R. Vaughn, R. Moorman, and D. Haines, "Frequency domain algorithm for quantifying atrial fibrillation organization to increase defibrillation efficiency," *IEEE Trans. Biomed. Eng.*, vol. 48, no. 9, pp. 969–978, 2001. DOI: 10.1109/10.942586

[31] T. Everett, R. Moorman, L. Kok, J. Akar, and D. Haines, "Assessment of global atrial fibrillation organization to optimize timing of atrial defibrillation," *Circulation*, vol. 103, pp. 2857–2861, 2001.

[32] M. Stridh, A. Bollmann, S. B. Olsson, and L. Sörnmo, "Time–frequency analysis of atrial tachyarrhythmias: Detection and feature extraction," *IEEE Eng. Med. Biol. Mag.*, vol. 25, pp. 31–39, 2006. DOI: 10.1109/EMB-M.2006.250506

[33] M. Stridh, L. Sörnmo, and S. B. Olsson, "ECG-based feature tracking in atrial tachyarrhythmias," *Comput. Cardiol.*, vol. 30, pp. 721–724, 2003. DOI: 10.1109/CIC.2003.1291257

[34] V. D. A. Corino, L. T. Mainardi, M. Stridh, and L. Sörmno, "Improved time–frequency analysis of atrial fibrillation signals using spectral modelling," *IEEE Trans. Biomed. Eng.*, vol. 56, 2009 (in press).

[35] F. Sandberg, M. Stridh, and L. Sörnmo, "Robust time–frequency analysis of atrial fibrillation using hidden Markov models," *IEEE Trans. Biomed. Eng.*, vol. 55, pp. 502–511, 2008. DOI: 10.1109/TBME.2007.905488

[36] L. Rabiner and B. H. Juang, "An introduction to hidden Markov models," *IEEE Acoust. Speech Signal Proc. Mag.*, pp. 4–19, 1986.

[37] G. D. Forney, "The Viterbi algorithm," *Proc. IEEE*, vol. 61, pp. 268–278, 1973. DOI: 10.1109/PROC.1973.9030

[38] A. L. Goldberger, L. A. Amaral, L. Glass, J. M. Hausdorff, P. C. Ivanov, R. G. Mark, J. E. Mietus, G. B. Moody, C. K. Peng, and H. E. Stanley, "PhysioBank, PhysioToolkit, and PhysioNet: Components of a new research resource for complex physiologic signals," *Circulation*, vol. 101, pp. E215–220, 2000.

Abbreviations

A/D	analog-to-digital (conversion)
ABS	average beat subtraction
ACG	atriocardiogram
AF	atrial fibrillation
ANOVA	analysis of variance
ANS	autonomic nervous system
APD	action potential duration
ApEn	approximate entropy
ARC	atrial rhythm classification
ASC	activation space constant
AT	atrial tachycardia
ATP	antitachycardia pacing
A:V	atrial to ventricular
AV	atrioventricular
BEM	boundary element method
bpm	beats per minute
BSPM	body surface potential mapping
CC	crosscorrelation
CCE	corrected conditional entropy
CD	correlation dimension
CE	conditional entropy
CHD	coronary heart disease
CHF	coronary heart failure
CRN	Courtemanche, Ramirez, Nattel (model)
CRT	cardiac resynchronization therapy
CS	coronary sinus
CV	conduction velocity
CVD	Choie–Williams distribution
DF	dominant frequency
ECG	electrocardiogram
EDL	equivalent double layer
EGM	electrogram
EMI	electromagnetic interference
ERP	effective refractory period

FFT	fast Fourier transform
FRP	functional refractory period
HMM	hidden Markov model
HRSH	heart rate stratified histogram
HRV	heart rate variability
ICA	independent component analysis
ICD	implantable cardioverter defibrillator
ILR	implanted loop recorder
LA	left atrium
LAW	local activation wave
LE	lower envelope
LP	level of predictability
LVH	left ventricular hypertrophy
MCE	mutual-conditional entropy
MR	magnetic resonance
MSC	magnitude-squared coherence
MSE	mean-square error
NLA	nonlinear association
NN	normal-to-normal
NSR	normal sinus rhythm
OACG	optimal atriocardiogram
PCA	principal component analysis
PDC	peak dominant change
PG	peak gap
pNN50	percentage of normal-to-normal RR intervals greater than 50 ms.
PV	pulmonary vein
PVR	peak value ratio
PWD	P wave duration
RA	right atrium
RMS	root mean-square
rMSSD	root mean-square difference of successive normal-to-normal RR intervals
SampEn	sample entropy
SAN	sinoatrial node
SDNN	standard deviation of normal-to-normal RR intervals
SE	Shannon entropy
SNR	signal-to-noise ratio
SOBI	second-order blind identification
SPA	second peak amplitude
SPP	second peak position

SR	sinus rhythm
STFT	short-term Fourier transform
SVD	singular value decomposition
SVT	supraventricular tachycardia
TMP	transmembrane potential
XWVD	cross Wigner–Ville distribution
VCG	vectorcardiogram
VT	ventricular tachycardia
WCT	Wilson central terminal
WVD	Wigner–Ville distribution

Index